ROTARY ELITE
ARMY CHOPPERS OF NATO

Yves Debay & James Hill

CONCORD
PUBLICATIONS COMPANY
CONCORD COLOR SERIES

by CONCORD PUBLICATIONS CO.
603-609 Castle Peak Road
Kong Nam Industrial Building
10/F, B1, Tsuen Wan
New Territories, Hong Kong

We welcome authors who can help expand our range of books. If you would like to submit material, please feel free to contact us.

We are always on the look-out for new, unpublished photos for this series.
If you have photos or slides or information you feel may be useful to future volumes, please send them to us for possible future publication.
Full photo credits will be given upon publication.

ISBN 962-361-725-4
Printed in Hong Kong

All photos by Yves Debay unless otherwise stated.

INTRODUCTION

As early as the sixteenth century, the famous inventor and artist Leonardo da Vinci made a drawing of a strange machine that could raise itself vertically into the air and be used for warfare. Nevertheless, it was not until 1930 that the first serious project to develop such a contraption was considered. Igor Sikorsky of Russia and a Spanish inventor named Juan de la Cierva were probably the first to build a flying machine like da Vinci had envisioned—which was not yet known as the helicopter.

The "Autogire" of La Cierva can be considered one of the first military helicopters, though it had a traditional engine and rotor. A few of these devices were built, and the RAF ordered one just before the outbreak of World War Two. At the end of the war, a Focke Wulf vertical takeoff machine was attached to a U-boat to be used for reconnaissance purposes, but the machine was still connected to the submarine with a cable.

The five years following the end of World War Two was the period that really saw the birth of the modern helicopter, with a dozen development projects being undertaken by the USA, the USSR, England and France. The first helicopters to be used for military purposes were flown by the US Army in Korea for liaison and small reconnaissance missions. In 1954, however, it was the Marines who launched the first troop-carrying helicopter assault.

The real pioneers of vertical takeoff assault were the French who, during the Algerian War, established the airmobile doctrine and tested it successfully on the field against guerrilla forces. A new soldier was born—the heliborne combatant—who arrived from the sky to fight on the ground. At the same time, the French used the first armed helicopter—a Sikorsky S-52 with a 20mm door gun.

In Vietnam, the American forces took airmobility to a new dimension with the creation of a full airmobile division, the famous First Cavalry, which was used in combat for the first time in the Ia Drang valley. With the ability at that time to mobilize more than 400 "choppers" at once, there was no doubt that the helicopter would make its mark in the history of warfare.

It was also in Vietnam that the first antitank helicopter was put to use. A Huey UH-1 was fitted with an early version of a TOW (Tube-Launched, Optically-tracked, Wire-guided) missile which was used against Viet Cong PT-76 and T-54 tanks during the 1968 Tet offensive.

In the '60s and early '70s, NATO nations had complete superiority in helicopters, but in the mid-'70s the Soviet Union started to develop the formidable Mi-24 "Hind", which can be considered a virtual flying tank. The Russians and their allies successfully used this fantastic machine in their wars of expansion in Angola, Ethiopia and Afghanistan.

With the threat of the Mi-24 and its successors, the Mi-28 "Hokum" and "Havoc", the Western countries were forced to develop a new kind of helicopter: the helicopter hunter. In fact, the first real combat between choppers took place in Iraq in 1980 when an Iranian Cobra shot down an Iraqi Gazelle (though back in 1973 an Israeli Super Frelon destroyed an Egyptian Mi-8 with a single burst of machine gun fire through the open lateral door). More recently, the Azeri used an Mi-24 to shoot down a gigantic Mi-26 during the fighting in Armenia.

Combat was not the original purpose of the helicopter. The first use of the new vertical takeoff weapon was for observation and liaison missions. A recce aircraft was not always available during the Korean and Algerian conflicts, so a regimental commander would often climb into a light helicopter to see for himself what was happening in the field. It was the era of the small Plexiglas-bubbled OH-47 and the French Allouette II.

The Vietnam War saw the arrival of the Hughes OH-6 "Loach", which was especially made to observe and direct fire. The reconnaissance helicopter was generally of a small size and was not armed. It was flown by a crew of two — the pilot and an observer — who had a good view and a sturdy pair of binoculars. The 1970s saw the introduction of a variety of observation and scouting helicopters, e.g., the OH-58, the Hughes 500, and the Gazelle.

The arrival of the antitank helicopter gives the observation helicopter a new reason for being. It has become a scout for the antitank helicopter, flying in advance of it to find and designate a target. It can also coordinate fire from aircraft like the A-10 or guide intelligent artillery projectiles. For these missions, the silhouette of the traditional recce helicopter would change. For example, the OH-58D is fitted with an observation sphere containing infra-red equipment, a laser rangefinder and a target designation system. It is probably guaranteed, however, that a strong pair of binoculars will still have its place in the cockpit.

In today's modern armies, the helicopter is omnipresent. These exceptional machines are used for countless tasks, including liaison and reconnaissance missions, artillery fire and transport, vertical assault, antitank and anti-helicopter missions, search and rescue, and electronic warfare. In only 30 years, the helicopter has become the key point of modern military strategy, and with more than 15,000 of the machines in its arsenal, NATO has become one of the biggest users. This book, then, is dedicated to the NATO chopper pilots who won the Cold War and will have an interesting future as NATO expands its airmobile and rapid deployment forces.

BELGIUM

The small country of Belgium, located in the middle of central Europe, is of great importance in relation to the communication in the immediate rear of NATO. Along with Rotterdam, the harbors at Antwerp and Zccbruggc arc thc main locations for landing US reinforcements in Europe.

At the peak of NATO's existence, Belgium had one army corps in Germany, which was made up of two divisions of two brigades each. Now the army has undergone restructuring, and all helicopter forces have left Germany and returned to Belgium to the operational base at Bierset. Belgium furnishes a paracommando brigade to the NATO's Immediate Reaction Force (IRF), with one helicopter battalion attached for antitank duties.

Belgium has had helicopters in its armed forces since 1958 when it received eighty SA 318 Allouette IIs. Delivered in batches over a period of time, they were used intensively by the Belgian Army for dropping commandos, artillery and aircraft fire control, medevac and liaison missions, and VIP transport duties.

Belgian Army's Aviation Group consists of two antitank helicopter battalions, a liaison helicopter battalion, and a maintenance company. Thirty-two Alouette IIs are still in service, serving alongside forty-six Agusta A 109BA (Belgian Army) helicopters. Twenty-eight are the "HA" antitank version fitted with the Saab/Emerson Electric HeliTOW antitank system, which includes 500 TOW missiles. Eighteen are the "HO" day and night reconnaissance observation version that are equipped with the Saab Helios observation system. If the need arises, the HA version can be quickly transformed to the HO version.

A Belgian Alouette II is shown here during a visit to an airfield in northern France. A prime example of a versatile rotary-wing aircraft, the Alouette II has been used extensively for both civil and military purposes. Note the square national flag that replaced the more traditional roundel.

This Alouette II of the Belgian Army is parked at the helicopter flight school at Brasschaat near Antwerp. Note the characteristic skid-type landing gear and open fuselage construction. Having the appearance of a bug's eye, the cabin of the Alouette is capable of seating a pilot and four passengers. Belgium will probably keep using this very capable helicopter for the next few years.

The third of the 46 Italian-made Agusta A 109 helicopters, whose delivery to Belgian Army was completed in early 1996, undergoes a maintenance check. Note the size of the main rotor, which measures 36 ft (11 m) in diameter.

Hovering over the airfield in Braschaat, one of the brand new A 109s (number "H02") prepares to land following a flight from Germany. In the background is A 109 "H04", which arrived at the same time. The A 109's sleek design is evident from this view.

This frontal view of A 109 "H04" shows the streamlined appearance of this model of helicopter. The first A 109s delivered to Belgium were the HO (observation) versions, which featured the Saab Helios roof-mounted observation sight that is seen so clearly in this photo. They can quickly be transformed into the HA (antitank) version, though.

The A 109 weighs in empty at 3,127 lbs (1,418 kg) and stands nearly 11 ft (3.30 m) high. It is able to maintain its slim silhouette because its landing gear is fully retractable, and its two Allison 250-C 20B turboshaft engines are located up inside the upper rear of its fuselage.

The A 109s acquired by the Belgian Army are painted in an overall dark green camouflage paint scheme, with the only color on the chopper being a small national roundel on the tailboom. This view of A 109 "H02" nicely displays the helicopter's 43-ft (13.05-m) length.

Exercise "Cold Grouse 95" in Denmark was the first operational sortie for the brand new Belgian Agusta A 109. The helicopters shown here, which are armed with TOW missiles, belong to the Belgian Army's 18th Antitank Battalion (18 Bn HATk). They take off from the airfield at Vordingborg in Zealand in Denmark.

Belgian A 109s touch down on open ground during "Cold Grouse 95". The Belgian Army Aviation Group has two antitank battalions, the 17th and 18th, which fly the A 109, and a liaison battalion, the 16th, that flies the Alouette II. Since the Belgian forces have left their garrison in Germany, the helicopters have been based in Bierset in Belgium.

The main armament of the A 109 consists of 8 TOW anti-armor missiles, which are fired by the Swedish-made HeliTOW system. On a normal mission, one company of Belgian A 109s is able to destroy one enemy tank battalion; a full battalion of them should be capable of destroying a regiment. The probability of hitting the target with these missiles is 90 per cent.

A trio of Belgian A 109 helicopters of the 18th Antitank Battalion engage in some formation flying over the Baltic Sea. A battalion consists of two antitank companies and one observation platoon, plus a staff and a maintenance and logistics company that are transported by wheeled vehicles.

During the "Cold Grouse 95" exercises in September 1995, a formation of Belgian Agusta A 109s practices antitank maneuvers. An antitank company consists of seven antitank helicopters and two observation helicopters. Note the national colors painted on the tails of these choppers.

Flying high above the lush landscape of Denmark, A 109s of Belgium's 18th Antitank Battalion train for an antitank attack. These Agusta A 109s have a maximum cruising speed of 177 mph (285 km/h) and a maximum range of 402 miles (648 km).

The profile view of this unarmed A 109, which belongs to the observation platoon of the 18th Antitank Battalion, was photographed during the "Cold Grouse 95" exercise. The battalion's observation platoon operates five A 109s. Note the Helios observation equipment on the roof of this helicopter.

A Belgian A 109 sends ripples across the top of a Danish pond as it hovers behind the camouflaging screen of a grove of trees. Ready to fire its TOW missiles, it waits in ambush for an unsuspecting tank. In the event of tension in Europe, the 18th Antitank Battalion would be attached to the NATO's Immediate Reaction Force.

A group of pilots from the 18th Antitank Battalion gather to compare notes following maneuvers during "Cold Grouse 95". These pilots are the best of any unit in the Belgian Army Aviation Group, and several of them have amassed more than 2,000 flight hours in a helicopter, especially the Alouette II. Their average age is 28-years-old.

This Beligan A 109 pilot is typical of the pilots in the Beligian Army Aviation Group. His shoulder patch shows the profile of an A 109 superimposed over the Belgian national roundel. The initials "BA" that follow the A 109 name directly above the roundel stand for Belgian Army. Note the fighting bird cartoon painted on the visor of his flight helmet.

As sunset approaches, two A 109s from the 18th Antitank Battalion return to base following maneuvers. These high speed and high performance helicopters are a noteworthy asset to NATO's European arsenal, as are the Belgian pilots who fly them.

DENMARK

Denmark is another small European country with great strategic importance. The Danish peninsula of Jutland and the island of Zealand, with its capital at Copenhagen, control the two strategic straits of Kattegat and Skagerrak . . . the only way by which the Russian Navy can leave the Baltic Sea. In case of trouble in the area, the Danish Army would have to be assisted by NATO reinforcements. So, every year a large NATO exercise takes place in that region.

Denmark has a small army air corps, the Haerens Flyvetjeneste, which consists of two squadrons based in Vandel in Jutland. Since 1975, the squadrons have been equipped with 14 OH-6 Cayuse liaison and reconnaissance helicopters. Only two OH-6s have been lost in more than fifteen years of service. Like the Americans who used them so extensively in Vietnam (and even gave them the nickname "Loach"), the Danes have been fully satisfied with the Cayuse.

In mid-1987, the Danish ordered from France twelve AS 550C2 Fennec helicopters, the military version of the AS 350L Ecureuil, equipped with the Swedish Saab/Emerson Electric HeliTOW antitank system. A new antitank squadron was formed with the twelve Fennecs, with the twelve OH-6 Cayuses used as scouts.

A Danish OH-6 Cayuse observation helicopter prepares to lift off from a makeshift airfield in Zealand during the exercise "Bold Grouse 88". The exercise was a combined Danish-British endeavor. In the background can be seen an RAF Chinook. Note the teardrop structure of the Cayuse's fuselage. The overall length of the fuselage and tail is 30 ft 3.75 in (9.24 m).

Manufactured by the American Hughes Helicopter Inc., the OH-6 Cayuse originally had the denomination of Hughes 500M. The Danish Army Air Service has two squadrons of helicopters in the Jutland Division. The first squadron only has one OH-6, but the second is an observation squadron with eight OH-6s. They are home-based at Vandel in Jutland.

A pair of OH-6 Cayuse light observation helicopters (LOHs, or loaches) painted in two-tone camouflage are seen here at their base at Slagelse during exercise "Action Express 93". Note the Danish roundel on the fuselage behind the cabin and the national flag guidon just forward of the yellow warning stripe on the tubular tailboom.

The Danish Army Air Service has six AS 350L Fennec helicopters (the military version of the AS 350 Ecureuil) in its antitank squadron. After being equipped with the Saab/Emerson Electric HeliTOW antitank system, they were redesignated AS 550C2s.

Sitting in the cabin of an AS 550C2 Fennec, this pilot of the Danish Army Air Service takes a break from his normal routine to pose for a photograph. The Fennecs participated for the first time in operational duties during the NATO exercise "Action Express 93" in which they were used to attack "enemy" armor with the Swedish-made HeliTOW system at a range of 4,098 yd (3,750 m).

The AS 350 (AS 550) was designed to be the successor of the Alouette, and the newer multi-purpose helicopter has proven to be every bit as reliable. The military version offers a variety of weaponry ranging from cannons and rockets to antitank and antishipping missiles. The AS 550C2 shown here displays the Helios observation system atop its cabin.

Each of Denmark's AS 550C2 Fennecs are powered by a 317 shp (236.5 kW) Allison T63-A-5A turboshaft, which provides the chopper with a cruising speed of 134 mph (216 km/h). Like other Danish helicopters, this TOW missile-toting AS 550C2 is painted in an effective camouflage paint scheme of black and dark green.

These two AS 550C2 were photographed during combat exercise "Action Express 93". Their camouflage paint schemes are effective against the dark green woods in the background. The Fennec in the foreground is easier to see due to the highly visible red protective tags on its rotors and the cover on its exhaust.

Its 7 ft 5.75 in (2.28m) long landing skids just barely clearing the ground, an AS 550C2 Fennec ascends skyward during military exercises. This profile view of the AS 550C2 allows us to get a good look at the overall 42 ft 5.5 in (12.94 m) length of the helicopter.

A squadron of Danish rotary-wing aircraft are lined up in a rural clearing for maneuvers during "Action Express 93". Note the Cayuse LOHs in the background. The main rotor of the AS 550C2, which measures 35 ft (10.7 m)in diameter, is clearly visible in this photo.

The overall height of the AS 550C2 Fennec is about 11 ft (3.3 m). Its tail rotor (which is nearly invisible as it rotates here) has a diameter of just over 6 ft (1.86 m). The yellow stripe at the end of the tail boom displays a red arrow pointing to the tail rotor and the words "DANGER KEEP AWAY".

A head-on view of a Danish AS 550C2 Fennec flying in an attack mode during "Action Express 93". Though seemingly weightless in this photo, the Fennec actually has a maximum takeoff weight of 4,960 lb (2,250 kg). When gassed up with 142.6 gal (540 l) of fuel, it has a range of 414 miles (666 km).

FRANCE

The French Air Force first used helicopters during the Indochina War (mainly for evacuation duties). In 1953 the Aviation Légere de l'Armée de Terre (ALAT), or Army Light Aviation was created with fixed wing light aircraft and some helicopters.

To counter the intensification of activity by guerrillas in Algeria, more helicopters were deployed, and a real doctrine of tactical mobility was evaluated. In 1956, the performance of helicopters such as the Sikorsky S-54 and S-55, Piaskey H-21 and the first Alouette II allowed the French staff to launch a large scale vertical encirclement and assault.

Ninety per cent of the operation in which the helicopters were involved was successful, and the first homogenous unit was created and named DIH (Détachement d'Intervention Héliporté). Using their Vertol H-21 "Banana", the DIH was able to transport an infantry company in only one rotation.

At the end of the Algerian War, the French developed a new vertical doctrine, making them the pioneers of the heliborne assault tactics that the Americans developed on a division scale with the 1st Airmobile Cavalry Division in Viet Nam.

The first armed helicopter was the H-19 "Chickasaw" (S-55) of the US Army and USMC, which was fitted with rockets and machine guns during the Korean War. The French were the first to introduce the antitank helicopter armed with wire-guided missiles: the Alouette II and III, which were equipped with SS-11 missiles. They saw action in Algeria, especially during action against the caves in the cliffs where some rebels were entrenched.

After the Algerian War, the size of the ALAT's helicopter unit grew, and some new units were created; they were named the RHC (Régiment d'Hélicoptere de Combat, or Combat Helicopter Regiment) and the GHL (Groupe d'Hélicoptere Lèger, or Light Helicopter Group). In the '70s, each of the French Army's three army corps contained two RHCs and one GHL. Some independent GHLs were assigned to the military territorial command.

The next step was the creation of an independent unit attached to the FAR (Force d'Action Rapide, or Rapid Reaction Force) that was totally airmobile. Expanding the concept of tactical mobility close to the front lines, the new unit, called the 4e DAM (Division Aéromobile), or 4th Airmobile Division), takes airmobile combat deep into the battlefield. After a flight of several hundred kilometers, the advance guard of the 4e DAM can resupply itself and launch an attack on a front 37 miles (60 km) wide and 62 miles (100 km) deep. On one pass, the 4e DAM can simultaneously launch 360 antitank missiles.

French helicopter pilots have a great deal of flight experience, especially at very low altitudes. All of the tactics of infiltration and strike were designed for extremely low levels. One French pilot says: "The low altitude means safety to us . . . Gazelles and Pumas don't have any protection, and in case of war we will fly very far behind enemy lines." French pilots acquired combat experience in Chad - where they used Gazelles to destroy Libyan vehicles - and during the Gulf War.

In case of war in central Europe, the 4^{e} DAM is the perfect tool for counterattack. Unlike the British, Germans, or even the Americans, who employ their combat helicopters defensively, close to the front in support of their ground troops, the 4^{e} DAM was created to strike at far distances. A Gazelle can make an initial strike after flying 124 miles (200 km) east, resupply at an airmobile post, picking up HOT missiles and fuel that were deployed in the field by a Puma, and make a new strike the same distance to the north.

Every year the 4^{e} DAM practices aircraft carrier landings on the Foch or Clemenceau. The warships are used as refueling platforms by a "box" of two Pumas and six Gazelles, which stay on board for only 20 minutes. Using this procedure, even if the ship is in the middle of the Mediterranean, the entire 4^{e} DAM can be deployed somewhere in North Africa in less than 24 hours.

Despite France's budget restrictions (the same problem that plagues the armies of other nations), the ALAT will maintain programs such as the Horizon (heliborne radar tested in the Gulf), the French-German HAC Tigre (antitank helicopter), and the HAP Gerfaut (support protection helicopter), which will provide the ALAT with the perfect tools to enter into the heliborne combat of the 21st century.

<u>ORDER OF BATTLE</u>

<u>The GHL</u>

The territorial command and each military area operates a GHL (Groupe d'Hélicoptere Léger, or Light Helicopter

An Alouette II belonging to the GHL of Nancy flies over the army camp at Suippes. This helicopter was used by the general in command of the 10th Armored Division to inspect his tanks. Production of the Alouette II ended in 1975, and with the production of more advanced helicopters, this model is not as commonly used as it was in previous years. Another version, the Alouette III, was built that incoporates a larger and more enclosed cabin and has a more powerful engine.

This AS 355 Ecureuil 2 (AS 555 Fennec) was used as a liaison during exercise "Farfadet 90". The ALAT acquired this twin-engined model of Fennec for training purposes at the EALAT (Ecole de l'ALAT, or ALAT School) in Le Luc in the south of France.

The French Army uses Gazelles mounted with cannon (SA 341F/Canon) and HOT antitank missiles (SA 341/342M). Here a flight of Gazelles from the 3e RHC armed with 20mm cannon demonstrate the might of the 4e DAM airmobile unit.

A French Gazelle pilot of the 3e RHC. These pilots are professionally competent soldiers. Some of them gained their combat experience during the war in the Persian Gulf, some in Chad fighting Libyan-backed guerrillas. Note how the large cabin door opens wide for easy access to the crew members' seats.

Gazelles of the 3e RHC touch down at a mobile refueling station in southern Corsica during the joint French-US exercise "Phyblex 89". The Gazelle's main three-blade rotor has a diameter of 34 ft 5.5 in (10.5 m). Its fenestron tail rotor is composed of thirteen alloy blades.

A 20mm cannon-mounted Gazelle of the 3e RHC which is painted in the desert camouflage scheme used in Chad. The 20mm cannon is normally used against soft vehicles and other helicopters. They are also used to cover the retreat of HOT-armed Gazelles after they have fired their last missile. Attached to the front of the cockpit are the twin dipole aerials of a radio homing device.

Gazelles of the 4e DAM depart Hyere airport for maritime flights during exercise "Damtam 90". The Gazelle's maximum cruising speed at sea level is 164 mph (264 km/h). It has a maximum range with standard fuel of 441 miles (710 km).

This 4e DAM Gazelle is armed with tube-launched, wire-guided HOT antitank missiles. It was photographed while rising skyward during the "Damtam 90" exercise. The Gazelle has a maximum rate of climb at sea level of 1,770 ft (540 m) per minute. Note the camouflaged Cougar (Super Puma) on the landing strip in the distance.

At a mobile refueling station in Corsica, a Gazelle of the 3e RHC takes on fuel during "Phyblex 89". In the background is a Puma transport helicopter.

With majestic, snow-covered mountains providing an impressive background, Gazelles of the 4e DAM arrive in Corsica to take on fuel after completing a maritime flight. Apart from its regular internal fuel tanks, the Gazelle has two extra tanks that can carry an additional 76.5 gal (290 liters).

Helicopters belonging to the 4e DAM await the arrival of their crews prior to embarking on the maritime flight during "Damtam 90". The open forward doors of the choppers create an interesting effect. Empty of crew, passengers and weapons, the Gazelle has a weight of 2,200 lb (997 kg).

Gazelles of the 4e DAM are shown here during a brief refueling stop aboard the aircraft carrier Clemenceau during the "Damtam 90" exercise. The choppers remained on the deck of the carrier for only 20 minutes.

Gazelles from the 4e DAM prepare to refuel in the shadow of a beautiful Corsican mountain range. While the overall height of the Gazelle exceeds that of its predecessor, the Alouette II, its overall length is almost exactly the same. The Gazelle is 10 ft 5.6 in (3.19 m) high and 39 ft 3.3 in (11.97 m) long.

An SA 341/342M Gazelle armed with HOT antitank missiles flies above an ERC-90 Sagaie armored vehicle. The helicopter, which belongs to the 5e RHC, was photographed while participating in the French-German airborne exercise "Kolibri".

Painted in brown and sand camouflage, and adorned with the French blue/white/red national roundel, a Gazelle sits at a barren airbase near the Iraqi border during the Gulf War. Both the forward doors and the smaller passenger/baggage compartment doors are open, giving a revealing look at the interior of the cabin.

An SA 341/342M Gazelle performs its primary function: the delivery of a Euromissile HOT antitank missile. The HOT missile has a range of 4,375 yd (4,000m) and can penetrate up to 31.5 in (800 mm) of solid armor. Mounted with the standard configuration of four HOT missiles, this Gazelle becomes a lethal rotary-wing weapon.

A participant in the "Farfadet 92" exercise, this Gazelle is painted in an unusual camouflage scheme of black, red-brown and olive drab. This profile view clearly shows a few of the familiar physical characteristics that identify the helicopter as a Gazelle: the enclosed fenestron tail rotor, the "raindrop" cockpit with its large doors, and the long, thin tailboom.

This close-up view of a Gazelle taking part in the exercise "Farfadet 92" shows the helicopter's three laminated glass-fibre rotor blades at rest. The white letter "N" is painted on the helicopter's battery access panel. Directly above the panel is the cabin ventilating air intake.

The French FAR (Force d'Action Rapide, or Rapid Action Force) employs a variety of helicopters in support of its operations. These three helicopters - two Gazelles and a Puma - belong to the 5^{e} RHC. They are soaring above some of the beautiful landscape located in Ploce, Croatia.

Members of the FAR helicopter detachment fly over a lake at Ploce in Croatia. The Gazelle has a surface ceiling of 13,450 ft (4,100 m). The two 5^{e} RHC Gazelles in the front of this formation are equipped with the Giat M.621 20mm cannon.

At the harbor at Ploce, where the FAR helicopters are based, it is common to see a mixture of French and British helicopters, such as the Gazelle and the Chinooks shown here. Note the "UN" identification markings worn by the helicopters serving in the Balkan war zone.

A ground crew at Ploce mounts a Giat M.261 20mm cannon to a Gazelle of the 5^{e} RHC. This weapon fires 3.53-oz (100-gram) projectiles at either 300 or 740 rounds per minute. Since the cannon only weighs 104 lb (47 kg) and has a light recoil, it is an ideal weapon for a helicopter.

The very sophisticated Eurocopter Tigre antitank helicopter will be the future combat helicopter of the French and German armies. It is equipped with a different rocket pod and the TRIGAT antitank missile, which is a "fire and forget"-type of armament. ALAT will take 215 of the new helicopter. (Eurocopter)

A French Army SA 330B Puma transport helicopter belonging to the 3e RHC flies over the deep blue water of the Mediterranean Sea during "Damtam 90", when it flew off the aircraft carrier Clemenceau. The original Puma became operational with the ALAT in 1970. An earlier model transport helicopter exists, the SA 321 Super Frelon, which is characterized by stabalizing floats attached to the support structure of its fuselage. However, the Cougar has since become the standard tactical transport helicopter of the French Army.

A close-up view of a French Puma shows the cockpit and cabin area. A few months after this photo was taken during "Damtam 90", this Puma was in the Persian Gulf (again on the Clemenceau) where it was stationed off the Saudi coast. Note the two-tone desert camouflage.

With its twin-wheel retractable tricycle landing gear lowered, a Puma lands on the carrier Clemenceau during the "Damtam 90" exercise. This hefty transport helicopter has a larger bulk and heavier weight than the lighter Gazelle. The Puma weighs 7,970 lb (3,615 kg).

Looking down the vast length of the flight deck of the carrier Clemenceau, two Puma helicopters can be seen being secured by the flight deck crew members following an exercise flight. At right is visible the conventional five-blade tail rotor of the transport helicopter.

The shadow of this Puma follows the helicopter on the desert floor as it flies near the Iraqi border during the Persian Gulf War. The white stripes that encircle the Puma's tailboom are a coalition force recognition sign. The maximum cruising speed of the Puma ("clean") is 168 mph (271 km/h). It seems odd that this helicopter is not painted in desert colors.

SA 330B Pumas of the 5e RHC engage in flight practice during an exercise called "Fregate". The Puma in the forefront of the photo still retains the two-tone desert camouflage it wore while serving in Chad. Note the air inlets into the two Turboméca Turmo turboshaft engines, each of which is rated at 1,575 shp (1,175 kW).

A large flight of Pumas belonging to the 5^{e} RHC based at Pau deploys in an open meadow during exercise "Fregate". During this exercise the helicopters practiced air assault techniques with the paratroopers of France's 11th Airborne Division.

A Puma of the 3^{e} RHC at Canjeurs participates in an exercise with the 1st Infantry Regiment, which is the infantry regiment operating within the 4^{e} DAM airmobile division. The RHC has two Pumas that offer logistic support to the Gazelles in the regiment. The Puma is capable of transporting the Jeep shown here by means of a sling.

An SA 330B Puma of the 5^{e} RHC that took part in the joint French-Spanish exercise "Iberia 91". The most noticeable characteristic of this particular helicopter has to be its paint scheme. This camouflage is quite a striking departure from the typical olive drab paint job.

An impressive array of French rotary-wing aircraft assembled as a regiment. An RHC has one squadron of Pumas and five of Gazelles. This aerial view shows an RHC gathered at Calvi in Corsica during an exercise.

With the details of their shapes blurred by their green/black/brown camouflage paint schemes, two Pumas from the 5^{e} RHC prepare to lift off into an overcast sky during the "Iberia 91" exercise. These versatile helicopters are capable of carrying 15 fully-equipped troops or 2 tons of cargo in their internal cargo holds.

France has a detachment of 6 Pumas called DETALAT, which is based at Djibouti. Here one of these Pumas flies over the volcanic "Devil's Island" in the Gulf of Tadjouhnah. Note how the landing gear is retracted into the fuselage during flight.

A Puma is seen here practicing the airlift of an M-2 howitzer. The helicopter's massive rotor blades create an infernal scene as it kicks up choking clouds of sand in the desert near Djibouti. The main rotor diameter of the Puma is 49 ft 2.5 in (15 m) and its tail rotor diameter is 9 ft 11.5 in (3.04 m).

An SA 330B Puma performs one of its foremost duties: transportation of equipment. In this case, it's an M-2 howitzer. The Puma can carry 2.5 tons of underslung material up to a range of 355 mi (572 km). Note the effectiveness of the camouflage painted on both the helicopter and its load.

This olive drab Puma is part of the "Iberia 93" exercise. It was captured on film as it lifted off from a field in the southwest part of France where the exercise was held. The maximum rate of climb of this helicopter is 1,810 ft (552 m) per minute. Its service ceiling is 19,685 ft (6,000 m).

Flight crews from the various Puma helicopters participating in "Iberia 93" gather for a briefing before take off. The pilots standing beside the Puma give us a good scale by which we can access the size of the Puma. It is 59 ft 6.5 in (18.15 m) long and 16 ft 10.5 in (5.14 m) high.

The successor to the SA 330 Puma is the AS 332 Super Puma which entered service in 1978. In 1990, the military variants of the helicopter were renamed Cougar and numbered AS 532. Here three Cougars from the 4e RHCM heavy support regiment rise skyward during the exercise "Farfadet 92".

One of the most prominent features that was new to the AS 532 Cougar is the nose radome that houses the optional Bendix/King RDR 1400 or Honeywell Primus 500 weather radar. This highly camouflaged Cougar is part of the 4e RHCM's contingent that took part in "Farfadet 92".

Although the Cougar (Super Puma) looks almost exactly like the regular Puma, there is another physical characteristic that distinguishes the Puma from the Cougar: the ventral fin located under the Cougar's tailboom below the fin.

These Cougars are part of the 4e RHCM based at Nancy. The rugged landscape over which they are flying is the Perpignan region of southeast France. Unlike the Puma, the Cougar is powered by two Turboméca Makila 1A1 engines that can rev up from an idle to full power in just 1.5 seconds.

The Cougar was built under license in Indonesia. Among the NATO countries, it is flown by Spain, Turkey and France. This AS 532 Cougar, which belongs to the 4e RHCM, is painted in the camouflage scheme of the Force d'Action Rapide. Note the large rotor head fairing on the rotor head; it is designed to smooth rough airflow that passes over the helicopter.

This rear view of a Cougar shows the scenic helicopter airfield in Perpignan in France where the Cougars that are part of the "Farfadet 92" exercise are secured after maneuvers are completed. The interior of the Cougar's fuselage can accommodate the crew plus 21 passengers. The development of a larger Cougar, the Cougar MkII (AS 332L2 Super Puma II), has increased that capacity to 28.

While an AS 532 Cougar sits on the tarmac at Perpignan at the time of "Farfadet 92", two Gazelles soar by in the sky in the distance. In 1991 this Cougar was one of six that were deployed to Turkish Kurdistan for Operation "Provide Comfort" where they flew humanitarian missions over Iraq to help the Kurdish population. It took 20 hours to cover the 2,608 mi (4,200 km) distance between France and Turkey.

A 3/4 view of one of the new NH Industries NH 90 helicopters. On its tail it displays the national flags of the four nations that participated in its production. France has put an order in for 220 of the NH 90s. The NH 90 is powered either by two Rolls-Royce/Turboméca/Piaggio/MTU RTM332 or two General Electric/Alfa Romeo T700 turboshaft engines. (NH Industries)

A Puma (right) and a Cougar speed through the air during the "Farfadet 92" exercise. With its increased load-carrying ability, the Cougar transport and assault helicopter has been much in demand as an export to countries such as Spain, Switzerland, Turkey, Singapore, Argentina and Brazil.

In cooperation with three other European countries - Germany, Italy and the Netherlands - France has become involved in the production of a new NATO helicopter, the NH 90 for assault transport, rescue, EW and VIP transport duties. This photo shows the front view of one of these new helicopters. (NH Industries)

GERMANY

Despite the ending of the Cold War and budget restrictions due to the reunification of the country, Germany remains the most powerful conventional force in western Europe with three army corps totaling twelve divisions. Though the recent changes will probably affect the number of divisions and the geographic location of the army corps, the helicopter forces should remain intact.

The Bundeswehr has more than 650 helicopters at its disposal, and each corps has its own Heeresfliegerkommando (Army Aviation Command). A small unit, Heeresfliegerbataillon 6, which has 21 BO 105 helicopters, is attached to the 6th Panzergrenadier Division, the unit assigned to the protection of the Schwlesig Holstein region in cooperation with the Danish Army.

Each Heeresfliegerkommando has 160 helicopters divided among three regiments and their support and logistics units. Among the regiments there are: one antitank regiment flying BO 105Ms and BO 105Ps; one light transport regiment flying UH-1Ds; and one medium transport regiment flying CH-53Gs.

The transport capacity of a Heeresfliegerkommando is either 1,300 men, 180 tons of material or 900 casualties. In only one rotation, a Heeresfliegerkommando is able to drop onto a battlefield a full paratrooper battalion and its support weapon.

For more than 30 years, German Army Aviation has been supporting the Bundeswehr and some other NATO forces in exercises involving all branches of the service and all weapons, as well as their daily training routine. The iron cross-marked Hueys are flying in countries as diverse as Norway, Denmark, Italy, and Turkey. They are often used to fly humanitarian missions and to assist the needy in the event of floods, forest fires, etc.

<u>The German Antitank Helicopter Unit Organization</u>

At corps level the antitank helicopters (ATH) are organized in regiments. At the division level (6th Panzergrenadier Division), there is one antitank helicopter squadron within Aviation Regiment 6.

These units are equipped with the PAH-1, an advanced configuration of the well known BO 105 liaison helicopter, equipped with HOT antitank guided missiles. Up to now it has not had any night combat capability, and adverse weather conditions (fog, poor visibility, low clouds or icy conditions) can limit or even prevent antitank operations. In general, though, this helicopter can be employed whenever weather permits the operation of armored formations over an extensive area.

Beside having command support, combat support, and security units, the typical ATH regiment is composed of two squadrons of four flights each. Each flight consists of seven ATHs. The ATH squadron of Aviation Regiment 6 has three flights. Consequently, there are 56 ATHs in each corps and 21 ATHs in the 6th Panzergrenadier Division.

<u>Missions and Operational Doctrine</u>

The mission of the ATH elements is to engage armored forces in all types of operations and in joint operation with ground forces. This means the integration of ATHs into the combined combat forces.

Because of the ATH's superior mobility and faster reaction time compared to any ground vehicle, the commander can quickly change their point of main thrust, disengage them (even during combat), and keep them in reserve.

The assembly area for an ATH regiment is a 25 sq mi (40-sq-km) area in the rear of the corps. For reaction times, the ATH regiment may establish up to three forward assembly areas in the division area. The forward assembly areas will be approximately 9.5 sq mi (15 sq km) in size and will consist of:

a command post;
dispersion sites for four flights;
and up to three forward area refueling and rearming points.

The supply installations of the forward assembly area are suitable for simultaneous HOT missile rearming and refueling of up to 12 ATHs within 10 minutes. Unlike the French and (in some cases) American helicopters, German helicopters that do not operate at the front lines are resupplied by trucks instead of helicopters.

If time permits, the ATH flight leader will give his mission briefing and present orders to the aircraft commanders at the command post of the forward assembly area. To save time, the orders may also be given at a rendezvous point near the area of operation.

Normally, the ATHs will be ready for action in their firing positions approximately twenty minutes after receiving the mission order.

Only France ordered more Alouette II helicopters than the German Air Force. Here is one of the two Alouette IIs that are still soldiering on in the liaison and observation flight of PanzerAbwehr (Antitank) Regiment (PzAbwRgt) 36 of Fritzlaar. Note the placement of the iron cross and German national flag on the fuselage below the engine.

Seen here is Alouette II number "7679", which was attached to the I GE Corps. This helicopter was used to transport VIPs during the exercise "Certain Shield". Two generations of helicopters are visible in this photo: the Alouette II in the foreground and the larger UH-60 Blackhawk.

A pair of Messerschmitt-Bölkow-Blohm (MBB) BO 105P (PAH-1) antitank helicopters belonging to PzAbwRgt 36 soar above the lush Hessian countryside. The two-tone black green/medium green camouflage is particularly well suited to the German landscape.

The first prototype of the BO 105 flew in February 1967 powered by two Allison 250-C18 turboshafts. In 1974 it was chosen to be the scout helicopter. In 1977 work began on the PAH-1 antitank variant, which was powered by twin Allison 250-C20B engines. One particular feature of the BO 105P is the APX M397 daylight-capable direct optical sight, located on the top of the main cabin.

An overhead view of a BO 105P antitank helicopter on the prowl in the skies over Germany. Visible on the port side of the fuselage are three of six Euromissile HOT antitank wire-guided missile system tubes configured to the antitank version of the BO 105.

The BO 105P's main rotor diameter is smaller compared to the French Puma or Gazelle; it is only 32 ft (9.48 m). At the time of this writing, MBB was just preparing for a two-phase PAH-1 upgrade program that includes retrofitting the helicopters with Allison 250-C20R3 engines and new rotor blades. The upgrade also includes HOT-2 missiles with a digital guidance system and the integration of a night vision system for both pilot and gunner.

A German BO 105P antitank helicopter belonging to the Fritzlaar-based PzAbwRgt 36 armed with six HOT missiles flies over a lake located near the what used to be the East German border. With a standard fuel load, the BO 105P has a maximum range of 408 mi (657 km).

This underside view of a BO 105P of PzAbwRgt 36, which shows a good view of its 38 ft 10.9-in (11.86-m) fuselage, was captured on film as the helicopter practiced low level tactical flying around Fritzlaar. The BO 105 is capable of great maneuverability; the pilot can practically waltz over the trees with his machine. The end plates affixed to the horizontal stabilizers carry HF/FM radio aerials for communication purposes.

Like a trio of giant locusts, these PzAbwRgt 16 BO 105P helicopters swarm over their assembly area during "Certain Shield 91". The BO 105 PAH-1 has a cruising speed of 144 mph (232 km/h). The next generation of combat helicopter to serve the French Army and the Bundeswehr will be the PAH-2 Tiger.

A BO 105P of PzAbwRgt 16 based in Celle waits for an "enemy" tank during the first great exercise of the new NATO Airmobile Division, "Certain Shield 91". This view offers a good look at the characteristic flat bubble front of the BO 105P's cockpit and its twin-skid landing gear. The height of this rotary-wing aircraft to the top of its rotor mast is 9 ft 10.1 in (3 m).

German pilots study their maps in preparation for a mission during "Certain Shield 91". Besides Germany, several other Western nations fly the BO 105s. The Netherlands has 27, which are used for recce and liaison missions. Sweden's BO 105s are armed with Saab/Emerson HeliTOW antitank missile system.

A BO 105P (PAH-1) takes on some much needed fuel from a Belgian fuel truck. The usable fuel load of the PAH-1 is 648 lb (294 kg); the total fuel capacity is 153 gal (580 liters). Note the interesting arrangement of the HOT antitank missiles on the fuselage pylon.

Under a joint agreement, France and Germany developed the Eurocopter helicopter designated as the HAC Tigre/HAP Gerfaut (antitank and escort versions) for France and the PAH-2 Tiger (antitank version) for Germany. This profile view of an PAH-2 shows the large diameter (42 ft 7.7-in [13-m]) main rotor slashing the air above the Tiger's 45 ft 11.25-in (14-m) fuselage. (Michael Jerchel)

A German PAH-2 Tiger puts on a demonstration of its aerodynamic abilities. The Tiger's climb rate is 1,970 ft/min (600 m/min). Its maximum cruising speed is 174 mph (280 km/h). Apparently, Germany is looking to order a total of 212 of these rotary-wing aircraft. (Michael Jerchel)

The stepped seating arrangement in the cockpit of the PAH-2 Tiger is reminiscent of other more newly designed helicopters. It allows for the fabrication of a more narrow fuselage. The presence of the people in the foreground helps us to better appreciate the 16 ft 6-in (5-m) height of the Tiger. (Michael Jerchel)

A close-up view of the fuselage of a Bell UH-1D helicopter of the 20 Heeresfliegertransportregiment (German Army Aviation Transport Regiment). This is one of 204 examples of the UH-1D that were ordered by the German Army's aviation command. The 5,210-lb (2,363-kg) UH-1D has a fuselage that is 41 ft 10.75 in (12.77 m) long.

This Heeresfliegerabteilung (Army Aviation Detachment) UH-1D, which was detached to the Allied Mobile Force, soars above a minaret and a mosque in Turkish Thrace during exercise "Archway Express 85". This variant of the Huey is powered by T53-L-11 engine, which is rated at 1,100 shp (820 kW).

During maneuvers in the Bavarian Alps in 1987, a UH-1D (Huey) flies in support of the 23rd German Mountain Brigade. The UH-1D's two-blade main rotor, which appears to be still in this photo, has a diameter of 48 ft (14.63 m). Note the Huey's distinctive short landing skids and "flat" cabin roof.

A UH-1D of the Heeresfliegerabteilung and a Royal Air Force Puma conduct an airmobile assault in support of Turkish paratroopers during "Archway Express 85" exercises in Turkish Thrace near the Bulgarian border. The UH-1D can carry one pilot and 12 troops in its cabin. As versatile as it is, the UH-1D may be replaced by the NH 90 multipurpose helicopter developed by France, Germany, the Netherlands and Italy.

French paratroopers of the 9e Régiment de Chasseurs Parachutiste deploy from a German CH-53G of Heeresfliegertransportregiment 25 to assault the Danube Bridge during the Franco-German exercise that was called "Kecker Spatz 88".

With its twin-wheeled main landing gears still protruding from the rear of its fuselage-mounted sponson fairings, a CH-53G transport helicopter ascends after delivering a cabin-load of French paratroopers during the "Kecker Spatz 88" exercise.

With its dark green paint job standing out boldly against the snowy background of the Alps, a CH-53G of Heeresfliegertransportregiment 25 stands ready to assist some German mountain troops during heliborne maneuvers. With its spacious box-like fuselage, the CH-53G can accommodate 55 fully equipped troops.

A CH-53G of Heeresfliegertransportregiment 25 joins in the action with German paratroopers during the Franco-German exercise "Kolibri 87" in Bade Watenburg. Note the cavernous appearance of the interior of the fuselage when the ramp is lowered.

Germany has 106 CH-53G transport helicopters in its aviation arsenal. Here number "80+70" is shown during a heliborne paratroop assault performed by elements of Fallshirmjägerbrigade 27 during the trial operation of the new NATO Airmobile Division during exercise "Certain Strike 92". The small armored vehicle, which is dwarfed by the 28 ft 5 in (8.66 m) tall CH-53G, is a Wiesel antitank vehicle equipped with TOW missiles.

During exercise "Certain Strike 92", a 35,000-lb (15,876-kg) CH-53G lifts off to perform more tasks. The white underside of the helicopter indicates that CH-53G "84+70" was used in Kurdistan for humanitarian missions.

The German participation in the first NATO Airmobile Division sortie was important; they provided more than 30 helicopters, including 25 CH-53Gs. This photo shows to advantage the CH-53G's six-blade main rotor and the four-blade tail rotor in action. Each of their diameters measures, respectively, 79 ft (24 m) and 20 ft (6.10 m).

This photo shows a line up of CH-53G transport helicopters at Vaerlos during the "Cold Grouse 95" exercise. In the new Bundeswehr organization, all of the transport helicopter regiments — 3 light regiments with UH-1Ds and 3 medium regiments with CH-53Gs — are regrouped into Heeresfliegerbrigade 3, which has its headquarters in Mending.

After having unloaded elements of the Dutch 11th Airmobile Brigade in the region of Stubbekobing in Zealand, Denmark, a German CH-53G rises up into the sky. The Bundeswehr will probably update its CH-53s and employ them into the next century.

Still wearing the white paint running the length of its underside that identifies this chopper as a veteran of the Kurdistan humanitarian mission, a CH-53G hovers before landing during maneuvers with the 11th Airmobile Brigade of the Royal Netherlands Army. The impressive size of the CH-53 is evident here. The overall length of this transport helicopter is about 99 ft (30.18m).

While ground troops disembark from the CH-53G in the foreground, two more of the transport helicopters heliport vehicles by means of a sling to an exercise field. The CH-53G derives its lifting power from the two T64-GE-16 turboshafts mounted on either side of the upper fuselage.

Germany is one of the countries that contributed to the development and production of the new NH 90 helicopter, providing 24 per cent of the work share. Eventually, Germany intends to enlarge its helicopter force by 272 of these NH 90s. This profile view provides an idea of how long the NH 90 is; it measures 63 ft 7 in (19.38 m) in length. (Michael Jerchel)

The two-man crew of this NH 90 takes their chopper skyward on a patrol. The hefty 12,566-lb (5,700-kg) NH 90 has an endurance time of 5 hours and 30 minutes. Two versions of the NH 90 were planned: the NFH 90, which is equipped with 360 degree search radar, and the TTH 90 (Tactical Transport Helicopter), which will perform assault transport, rescue, EW and VIP transport duties. (NH Industries)

This in-flight photograph of the new NH 90 medium-lift helicopter shows off the sleek appearance of the chopper after its landing gear has been retracted. Note the length of the rotor blades. The main rotor, whose four blades were manufactured using new techniques involving resin components, has a diameter of 53 ft 5.5 in (16.3 m). (Michael Jerchel)

GREECE

Located at the southern point of the Balkan peninsula, the small, sunny country of Greece, which has long been a major site for tourism, is of great strategic importance in the Mediterranean Sea.

Greece was often considered the "problem child" of NATO, having left the alliance (rejoined in 1980) after the Turkish invasion of Cyprus when the USA maintained a benevolent neutrality in favor of Turkey. The Turkish problem is still not solved, the northern part of Cyprus still being under the control of the Turkish Army.

For all these reasons, Greece still retains a heavy defense budget of 6% of its gross national product and a strong armed force of 170,800 men for a population of 10,500,000, consisting of four army corps. Greece has a small Army Aviation branch (Elliniki Aeroporia Stratou) made up of battalions called "tagma".

The first battalion, or 1 Tagma Elliniki Aeroporia Stratou (TEAS), is based at Stefanoviklio near Larissa in the center of the country. The second, 2 TEAS, is based at Megara near Athens while the third, 3 TEAS, is based at Alexandroupoli.

The 12 newly acquired AH-64A Apaches equipped the 1 Tagma Epidolkon ElikoPteron (1st TEEP - 1st Attack Helicopter Regiment), based at Stefanoviklio. The Hellenic Army Aviation has interest to acquire an additional 12 AH-64As.

The helicopters of Hellenic Army Aviation are meant to handle transport and heliborne assault and support duties in the event of a crisis in the Aegean islands. Hellenic Army Aviation operates 12 AH-64As, 76 UH-1Hs, 50 AB 205s, 12 UH-60As, 16 Bell 206Bs, 9 CH-47Ds, 1 AB 212 and 15 NH 350Cs. Greek helicopter pilots are very professional; fifty per cent of them have more than 2,000 flight hours. They also have the distinct advantage of flying over a country that is not very large and that they know very well. One Greek pilot muses, "If we are lost, we open the door and, by smelling the herbs in the mountains, we know where we are."

In short, Hellenic Army Aviation may be small, but its force is efficient, flexible, and lead by competent and motivated officers.

This Bell UH-1H flown by the Hellenic Army Aviation (Elliniki Aeroporia Stratou) was used for VIP transport. Every battalion (*tagma*, or TEAS) in the aviation branch utilizes the versatile UH-1H. They also use the Agusta-Bell AB 205, which is identical in appearance to the Huey. The Greek Army even has a training school (Skoli Aeroporias Stratou) where Hueys, AB 206Bs and Nardi-Hughes NH-350Cs (which have a "bug-eye" cabin like the Alouette II) are used to prepare the students for helicopter service.

Soaring high above some of the typically scenic countryside of Thessalonica in northeastern Greece, two UH-1Hs from 1 TEAS practice some formation flying not far from their base at Stefanoviklio. Note the overall dark green paint job they sport.

A UH-1H of the 1st Light Aviation Battalion based in Stefanoviklio. Suspended from the transport sling under the helicopter is a crate filled with fuel containers. Hueys are divided between the first and second *lokos* (company) of each battalion.

These Hellenic Army Aviation UH-1Hs from 1 TEAS nearly blend in with the mountainous terrain of Thessalonica. The helicopters belong to either the first or second company of the battalion. Number 3 Lokos flies AB 206Bs for artillery-spotting and observation.

Ground crew members prepare a UH-1H for flight on an airstrip in Greece. Note the flexibility of the main rotor blades as they angle downward when the red safety ribbon is removed. This angle provides a clear view of the compact powerplant housing on top of the cabin roof.

Parked on the ramp at their home base at Megara are several elaborately camouflaged CH-47C helicopters made by the Italian company Elicotteri Meridionali. These particular medium-lift helicopters belong to 2 TEAS. Their main duties involve the transporting of paratroops and the supplying of army units. The CH-47Cs has now been upgraded to CH-47D standard by Boeing.

This formation of UH-1Hs from 1 TEAS represents a small percentage of the 76 Hueys flown by the Hellenic Army aviation branch. While the UH-1Hs have been the workhorses of the army aviation, it became necessary to order AH-64A Apache antitank helicopters to strengthen the inventory of each battalion.

ITALY

With a population of 57.7 million inhabitants and a strategic position in the middle of the Mediterranean, Italy is still one of the most important NATO countries. Italy not only has a brilliant historic past but it can be considered an unsinkable aircraft carrier on the southern flank of the NATO alliance. With the end of the Cold War and the political wish to reorient the focus of NATO to the south, its strategic importance can only grow.

Origins of the aviation component of the Esercito Italiano (Italian Army) can be traced to 1911 when, for the first time, aircraft were used to observe and attack the Turkish enemy in Libya. During the following year, and in World Wars One and Two, all the flying machines were attached to the Air Force (Regia Aeronautica), and it was only in 1951 that the Esercito decided to create an autonomous light aircraft force with some Stinson L-5s performing observation duties. Despite a conflict with the Air Force, some Piper Cubs were delivered, and in 1956 the first helicopter operation began using the Agusta-Bell 47G. But it was only on December 31, 1959 that the Army Light Aviation was created and the name Aviazione Leggera dell'Esercito (ALE) was officially adopted.

Machines and Missions

Observation and reconnaissance are the ALE's main tasks, and all the types identified by the ALE's own classification system with the abbreviation "ERI" (Elicoterro da Ricagnizione, or recce helicopter) are devoted to these missions, which also include target acquisition. The most widely used machine for such a mission is the Agusta-Bell AB 206C-1 of which about 110 examples remain in service out of the 136 delivered since 1970.

Air mobility is the ALE's other main task. It is carried out by its fleet of EMs (Elicottero Multiruolo, or multirole helicopters) and ETMs (Elicoterro de Trasporto Medio, or medium transport helicopters), the latter being the designation for the Elicotteri Meridionali-Boeing CH-47C Chinook built under license in Italy, which ranks third in the world in the exportation of that helicopter.

Of the Chinooks received, only two have been lost since deliveries began in 1973, one of them during a demonstration tour in the United Arab Emirates. Recently three machines were delivered to resupply the fleet. All of the 38 Italian Chinooks will be brought up to C/D standards with more powerful engines and new rotor blades and avionics.

The bulk of the transport force is still made up of 92 AB 205As or EM-2s, the Italian version of the famous UH-1H. This ubiquitous helicopter has the capacity to carry nine soldiers, but in the service of the ALE it is used as a true multi-role machine thanks to the versatility of its design. Although the oldest helicopter in service, the AB 205A is very popular with its crew, who wish only for new radio equipment and armor plating.

The ALE has sent a detachment of AB 205As on peacekeeping missions to Namibia, Lebanon, Kurdistan, and the former Yugoslavia. The machine has paid a heavy price for peace - one AB 205A was shot down in Lebanon and one AB 205A was destroyed in the air by a federal MiG-21 in Yugoslavia, killing 6 EEC observers.

The Agusta-Bell AB 212/EM-3 was ordered as the successor to the AB 205A, and deliveries began in 1983. However, it proved less than satisfactory in ALE

An AB 206C-1 scout helicopter of 4th Reggimento ALE "Altair" flies over the Italian Alps during the exercise "Display Determination 86". A licensed version of the Bell Model 206A, the AB 206C-1 has its origins in the earlier UH-1 Iroquois (Huey) helicopter. The evolution of design from the Huey to the AB 206 is most noticeable in the shape of the cockpit.

AB 206C-1 general purpose light helicopters at the ALE flight school in Viterbo. The AB 206 is generally smaller in size than the earlier AB 204 (UH-1). Its rotor diameter, its overall length and height, and its weight are all less than those of its predecessor.

A classic example of the AB 206C-1 used by the Esercito Italiano (Italian Army). A close examination of the helicopter will reveal a subtle black green/dark green camouflage scheme on the cabin and fuselage. The Italian roundel, the warning arrow on the tailboom, and the stripes on the rotor tail provide the only color, beside the white aircraft numbers, identification name and landing skids. (Eagle Press)

service and only 19 remains in service. A more powerful replacement for the AB 205A appears to be the AB 412 or EM-4. Deliveries began in 1986 despite financial problems, and at current time the number of machines that has reached operational units totals 23. While the type's transport capacity is almost identical to that of the AB 205A and 212, the 412 offers better general flying characteristics and is more powerful.

One further type has recently been charged with non-operational transport missions and liaison flights: the Agusta A 109EOA or EC-1 (Elicoterro de Collegamento, or liaison helicopter). This acquisition was related to trials conducted to evaluate tactics and weapon systems for the A 129, but when examples of this attack helicopter became available for the operational evaluation phase, the A 109EOA was no longer required.

A major and more active part in Italian military operations is foreseen for ALE in the future, and this is due particularly to the introduction of the Agusta A 129 Mangusta, classified as EA, or Elicotero d'attacco (attack helicopter), which is now in use in a special attack regiment. This gives to the ALE a true night and day

One of the A 109EOA light helicopters that served as liaison and reconnaissance helicopter. The A 109EOA is recognizable by its streamlined appearance, which exaggerates its 35-ft (10.7-m) length and minimizes its 10 ft 10-in (3.3-m) height. It is a versatile chopper that can be fitted with an arsenal ranging from machine guns to rocket pods.

One of the first five Agusta A 129s in the Italian Army hovers over the tarmac at the ALE school in Viterbo. This profile view shows the offset positioning of the pilot and co-pilot/gunner in the narrow cockpit. The 40 ft 3 in (12.3 m) length of the A 129 is well exhibited in this photograph.

This view of the Italian Army Agusta A 109EOA shows the retractable landing gear to advantage. The A 109EOA weighs 3,332 lb (1,466 kg). Its two 420 hp Allison 250-C20B turboshaft engines give it a maximum cruising speed of 169 mph (272 km/h). This example is now used as a liaison and reconnaissance helicopter.

The Agusta A 129 Mangusta (Mongoose) is the first Italian Army antitank helicopter. The A 129 can be armed with a variety of weapons. This one is fitted with two TOW missile launchers and two seven-shot rocket pods. The Mangusta could also carry HOT or Hellfire missiles. Conspicuously located on the nose of the chopper is the TOW telescopic sight, with a laser and FLIR.

This rear view of an A 129 Mangusta in flight at the ALE flight school at Viterbo allows us to size up the height of the helicopter, which is 10 ft 11.9 in (3.35 m). Extending horizontally from the fuselage and from the rear of the tailboom are the stub wings and the tailplanes.

The A 129 Mangusta has a powerplant of two Rolls-Royce Gem 2 Mk 1004D turboshaft engines to power its four-blade main rotor, which has a diameter of 39 ft (11.9 m). Because of the similarity in its design to the McDonnell Douglas AH-64, the A 129 can justifiably be called the "mini-Apache".

A close-up view of the cockpit of an A 129 Mangusta places its pilot and co-pilot/gunner in the limelight. The stepped seating design, which puts the pilot in the elevated rear position, provides the co-pilot/gunner with a clear field of view for more accurate operation of weapons.

antitank capability. The Mangusta is a dedicated tank buster and is a part of the new Catrin C-31 system under deployment by the Italian Army. The first series of machines was delivered in 1992 and deliveries have continued since then.

The ALE is also interested in the scout and escort version of the A 129 armed with air-to-air missiles and a gun. The ALE introduced 24 Agusta A 109EOAs, or Elicotero d' Osservazione Avanzata (advanced observation helicopter). They were given to some squadrons that formerly operated the AB 206C-1.

At the beginning of the next century ALE will have a force of 450 helicopters of different classes: light helicopters represented by the A 129 in its different versions, medium helicopters (the NH 90); and heavy helicopters (the upgraded CH-47).

Organization

The basic flying element of the ALE is the *squadrone* (squadron), which has a statutory strength of six helicopters or four aircraft. Two or more squadrons form a *gruppo squadroni* (squadron group), which also comprises service and logistics elements. Two or more *gruppo squadroni* form a *raggruppamento* (wing), and all of their names relate to stars or constellations.

The ALE was restructured in 1996, and it is basically comprised as follows:

The 1st Reggimento "Antares" based at Viterbo in central Italy. It operates some Chinooks and will soon be issued some Piaggio P 180 Avanti transports for military use;

A new multi-role regiment, the 2nd Reggimento "Sirio", is to be established. It will be based at Lamezia Terme in the south of Italy. Its task will be to provide airmobility to the land forces;

An ALE *commando*, or command attached to the 3 Corps in charge of the defense of northwestern Italy has been upgraded as the 3rd Reggimento "Aldebaran";

The 4th Reggimento "Altair" reports to the 4 Corps of "Alpine" troops that operate from the border of France to that of the former Yugoslavia;

The 5th Reggimento "Rigel" is assigned to the 5 Corps. The largest field army unit, it is charged with the defense of the eastern part of northern Italy.

In addition to the regiments, there are seven autonomous units, five of which report to the local territorial military command. One unit supports the needs of the Folgore paratroop brigade, and at least one, the 39th Gruppo Squadroni "Drago", is charged with special operations tasks under the direct order of the Minister of Defense.

Two specific liaison flights have been formed and assigned to the 28th Gruppo Squadroni at Rome and the 53rd Gruppo Squadroni at Milan. They operate, respectively, A 109EOAs and AB 412s.

While still parked on solid ground, the crew of this A 129 Mangusta goes through a succession of routine pre-flight checks before departing their base. Note how the A 129's cockpit side windows are set in doors that are hinged to open upwards and the "Agusta A 129" name lettered on the engine casing. (Eagle Press)

The missile launchers and rocket pods fitted to this A 129 give the helicopter a far bulkier appearance than its usual narrow width. When loaded down, the weight of the Mangusta increases from its empty weight of 5,575 lb (2,529 kg) to 8,157 lb (3,700 kg). (Eagle Press)

An overhead view of an A 129 Mangusta in flight. The Mangusta's twin turboshaft engines allow the chopper to obtain a cruising speed of 155 mph (250 km/h) and a maximum speed of 168 mph (270 km/h). Fitted with an antitank armament, the A 129 has an endurance time of 2 hours 30 minutes. (Eagle Press)

This photograph of a lethal A 129 of the Esercito Italiano was taken in Somalia during Operation "Restore Hope". This 3/4 view of the tail of the helicopter shows off the two engine casings that are fitted to each side of its upper fuselage. (Eagle Press)

The hot Somali sun reflects off the rear fuselage of A 129 number "E.I. 917" as casually dressed crew members hustle around the chopper during Operation "Restore Hope". Note the manner in which the rocket pods and missile launchers are attached to the port wing pylon. (Eagle Press)

With its five-bladed main rotor churning the air, a special scout variant of the Agusta A 129, the International, hovers in the air as it displays its deadly nose-mounted 20mm rotating gun. This particular example is also fitted with 19-round rocket pods and Stinger missile launchers. It is powered by LHTEC T800 engines. (Eurocopter)

In a scene that is eerily reminiscent of an airmobile operation during the Viet Nam War, but which was captured on film during the exercise "Allegro Exchange 86", an AB 205A from the 5th Rgt ALE "Rigel" unloads soldiers from the AMF combined infantry unit that included paratroops from Belgium, Germany and the US.

One of the 92 AB 205As that the Italian Army uses for transport duties. Both the sliding doors on each side of the cabin have been opened to show us the seating arrangement and storage capacity inside. Note the A 129 Mangusta parked in the background. (Eagle Press)

During the NATO exercise "Allegro Exchange 86", Italian AB 205As from the 5th Reggimento ALE "Rigel" transport infantry during airmobile maneuvers. Visible in this photo is the small stabilizing bar that is attached at right angles to the lengthy main rotor blades to help balance the lifting rotor.

Freshly painted and pristine in appearance, AB 205A number "265" of the 4th Rgt ALE "Altair" fits in well with the magnificent surroundings of northern Italy. This AB 205A is armed for combat with twin mounted MG3 machine guns at the doors of the rear fuselage.

This close-up photo is a study of some of the weapons that can be fitted to the AB 205A. At the bottom is a seven-tube 2.75-in (7-cm) rocket launcher. Directly above this is the M134 7.62mm Minigun fitted to a gun mount attached to an external stores pylon. The small gun drive motor is closest to the camera. (Eagle Press)

A nice profile shot of an AB 212 multirole helicopter belonging to 30th Gruppo Squadroni "Pegaso", one of Italy's independent units. It is shown parked at the ALE school in Viterbo. Note how the red, white and green colors of the Italian flag have been painted on the tips of the tail rotor blades.

Dressed to beat the heat in Somalia, a ground crew member cleans the door-mounted MG3 machine gun on an AB 205A during Operation "Restore Hope". Sometimes it's easy to forget that none of the AB 205As in the Esercito Italiano would get off the ground if it weren't for the diligent service of their ground crews. (Eagle Press)

An AB 212 of 30th Gr Sq "Pegaso". The AB 212 is the Italian licensed version of the twin-turbine Bell Model 212/UH-1N. While it resembles the AB 205 somewhat, it is longer and larger overall. Its main rotor diameter is 48 ft 2.25 in (14.7 m) and it is 57 ft (17 5 m) long and 14 ft 10.25 in (4.5 m) high.

The AB 412 is the Italian version of the Bell Model 412, the advanced technology variant of the AB 212. Its main physical difference from the AB 212 is its four-blade main rotor. A few of the ALE's regiments use the AB 412. The example shown here is equipped with a 20-round rocket pod. The first flight of a Model 412 was made on August 3, 1979.

In Italy the 412 military version of the "Huey" is built as the "Grifone". The AB 412 is an improved version of the Model 212, the civil version of the UH-1N variant. It is faster and flies more quietly than the earlier variant, and it has an increased payload. Its maximum cruising speed at sea level is 143 mph (230 km/h).

This photo of AB 412 number "466" was taken just after the chopper touched down for refueling. The four-blade main rotor has slowed, but the tail root is still blurry from rotating. The AB 412 is a substantial helicopter, its weight being 6,470 lb (2,935 kg).

Crew members inside the cabin of this Italian AB 412 helicopter watch as French soldiers make preparations to refuel the helicopter during joint maneuvers. The AB 412 has an internal fuel capacity of 330 gal (1,249 liters). This profile view of the fuselage gives us a good look at the characteristic cowling structure and the subtle camouflage pattern.

The view from underneath this AB 412 shows the 46-ft (14-m) four-blade main rotor to advantage. This new example belongs to Gruppo Squadroni "Leone" of the 1st Rgt ALE "Antares". Note what appears to be a rescue hoist/winch affixed above the starboard cabin door.

CH-47C number "E.I. 817" of 11th Gr Sq "Ercole", with its rear loading ramp lowered, appears ready for inspection on a drill field in the shadow of the impressive Alps. Note the effectiveness of the helicopter's gray/green camouflage scheme in the surrounding environment.

The CH-47C is powered by T55-L-11A turboshaft engines mounted in pods attached to the rear motor mast. The two rotor masts feature two three-blade rotors that are 60 ft (18.3 m) in diameter. They help to give the Chinook its impressive size - 98 ft 10.75 (30.1 m) long and nearly 19 ft (5.7 m) high.

A CH-47C Chinook medium-lift helicopter belonging to the 1st Rgt ALE "Antares" soars high above the rugged terrain of Sardaigna during the exercise "Dragoon Hammer 90". The CH-47 can attain a cruising speed of 159 mph (256 km/h) at optimum altitude.

One characteristic feature of the CH-47 Chinook is the series of five circular cabin windows along the length of its fuselage. Here a 23,093-lb (10,475-kg) CH-47C, with a French Gazelle as an escort, is shown parked in one of nature's natural airfields. The Chinook rests upon its quadricycle landing gear. The main landing gear in front are twin-wheel units while those in the rear are single.

A machine gun mounted in the doorway of the main entry door of this Italian Army CH-47C provides the transport helicopter with some measure of protection. This close-up shows the profile of the nose of the CH-47C that is so prominent. (Eagle Press)

Over the next several years, the Italian Army will be phasing out the use of the AB 205A, AB 212, and the AB 412 for medium-lift duties. They will be replaced by the NH 90, such as the one shown here, which Italy helped to make a reality. (NH Industries)

The impressive size of the NH 90 - it is 17 ft 9.5 in (5.4 m) high - is evident in this photo. The cabin of the NH 90 is large enough to carry 20 fully-equipped troops up to a range of 1,553 miles (2,500 km). Italy's ALE intends to eventually amass a total of 214 NH 90s in its helicopter arsenal. (NH Industries)

NETHERLANDS

Located in the central part of northern Europe on the borders of Belgium and Germany, the Netherlands is one of the richest and industrialized, but smallest, countries in NATO.

The Dutch Army has only one corps, the 1 NL Corps, and half of it is based in Germany. Changes are now being carried out to disband the 1 NL Corps and two of its three divisions, the remaining division being integrated into the Dutch-German Corps. The 11th Airmobile Brigade is newly formed and can be assigned to different NATO operations.

The Light Aviation Group, which is normally located at Deelen, draws its resources from the Royal Netherlands Air Force (Klu), which recently began expansion from three to five squadrons.

The Order of Battle is as follows:

298 and 300 Squadrons are equipped with SA 316 Alouette IIIs and CH-47Ds. 300 Squadron received its first AS 532U2 in April 1996;

299 Squadron is equipped with the BO 105CBs;

302 Squadron was formed within the existing 300 Squadron. It was moved from Deelen to Gilze-Rijen and changed from a training unit to a combat unit;

301 Squadron is another new combat unit that was formed at Gilze-Rijen. This squadron operates the AH-64D Apache which is on loan until the delivery of the 30 AH-64 on order.

After the French Sud Aviation Alouette III was selected for use, 77 were delivered to the group from 1964 on. There are currently 58 of them still in service.

In 1975 a new helicopter rounded out the Royal Netherlands Air Force Light Aviation Group, the BO 105CB, 29 of which entered service. The Dutch BO 105s were recently fitted with night vision instruments and its crew members were issued night vision goggles.

The Dutch Light Aviation Group will be expanded to a force of 25 transport helicopters, 40 armed helicopters and the 27 BO 105CBs. The Netherlands took delivery of the first of 17 AS 532U2 Cougar transport helicopters in April 1996, equipping the 300 Squadron. While it was originally believed that the Dutch government would select the Eurocopter Tiger as their new attack helicopter, they actually selected the AH-64D Apache, placing an order in December 1995 for 30 of these choppers. The first of 13 CH-47Ds arrived in the Netherlands in December 1995. They are used to fly support missions for the 11th Airmobile Brigade.

Alouette III number "A-342", which belongs to 298 Squadron of the Royal Netherlands Air Force (KLu), hovers above a wargame airfield during the exercise "Certain Shield 91" in which the Alouettes simulated antitank helicopters. The Dutch armed forces are one of the primary users of the Alouette III, though under the new reorganization this squadron's Alouette IIIs will be replaced with transport helicopters. Other NATO countries that fly the Alouette III are Spain, France and Portugal.

With his foot resting on the starboard boarding step of his Alouette III, a Dutch pilot from 298 Squadron of the Royal Netherlands Air Force makes final preparations for a mission. The Alouette III made its first flight on February 28 1959. Despite its age, the Alouette III was popular among the pilots that flew it during the "Certain Shield 91" exercise.

This close-up view of a 298 Squadron Alouette III shows off the many curved windscreen panels that encase the helicopter's cockpit. Note how the crudely painted camouflage scheme partially obliterates the tricolor national roundel at the rear of the fuselage. The cabin of the Alouette III can accommodate two crew members and up to five passengers.

A black and green camouflaged BO 105CB observation helicopter surveys the action at the French military camp at Mourmelon, where the Dutch 13th Mechanized Brigade is allowed to conduct exercises. The agile and aerobatic BO 105CB is propelled by a four-blade main rotor with a diameter of 32 ft 3.5 in (9.84 m).

The Royal Netherlands Air Force Light Aviation Group contains a certain number of female pilots, like the one shown here in front of her BO 105CB. She is taking a break from the action during "Certain Shield 91" that was held in Germany.

Several of the BO 105CB helicopters that Dutch pilots flew during the exercise "Certain Shield 91". These BO 105CBs, which flesh out the larger helicopter force of Alouette IIIs, belong to 299 Squadron based at Deelen (later based at Gilze). They are each powered by two Allison 250-C20B turboshafts that are rated at 420 shp (313 kw).

NORWAY

Situated in northern Europe, Norway is very strategically positioned because the country controls the northern access to the free seaways used by the former Soviet fleet based at Murmansk and Arghangelsk, the two main ports on the Arctic Ocean. Norway also controls the Baltic straits of Kattegat and Skagerrak. Given its oblong length and small population of 4.3 million, Norway is a challenge to defend. In case of war, NATO reinforcement would be vital.

Helicopters play a major role in the country of Norway, especially since there are few roads. Fjords and mountains covered with snow nine months of the year are the characteristic features of its geography. But it is an excellent terrain for defense, and Norwegian troops can be deployed on the mountain peaks by helicopters.

The helicopter force of Norway is a part of the Royal Norwegian Air Force. It consists of two squadrons which are: Squadron 339 based at Bardufoss in the north and Squadron 720 based in Rygge in the south. The helicopter force was formerly equipped with UH-1Bs, now being replaced by Bell 412SPs. The first Bell 412SP arrived in Norway in 1987, others were assembled in country with US spare parts. They can transport nine soldiers equipped with skis and have been fitted with radar warning systems and chaff dispensers.

The Bell UH-1Bs used by the Norwegian Army normally perform traditional transport and assault missions in conjunction with the 6th Norwegian Division, but they can also assist civilians during hard winters. The UH-1B may be called upon to take fresh feed to reindeer or supply food to an isolated farmer.

This Bell UH-1B, which belongs to Squadron 720 from Rygge in southern Norway, made a landing in a forest near Oslo during a liaison mission. This photograph was taken during the NATO exercise "Blue Fox" in September 1988.

A Bell UH-1B of Squadron 339 soars high above the snowy grandeur of Norway. This machine is seen during a recce mission that was part of the NATO exercise "Arrowhead Express" in March 1988. Note the rescue winch attached to the starboard side of the cabin.

A brand new Bell 412SP from Squadron 339 is refueled at a makeshift airfield during the exercise "Teamwork 92". The 412s replaced the older UH-1Bs. Note the various colors of the three-tone camouflage. While probably quite effective in the forest regions, it is useless in the stark white of the snow.

Bell's Model 412SP features an increased gross weight and fuel capacity. This model, along with the 412HP, accounts for most of the military sales of this helicopter. This frontal view of the helicopter shows to advantage the chopper's characteristically wide fuselage.

During the "Teamwork 92" exercise, a Bell 412SP floats in front of a background of snow-capped and tree-covered mountains. The first Bell 412SP that arrived Norway crashed due to a mechanical failure. Norway builds its own Model 412SPs at Helikopter A/S at Stavenger.

PORTUGAL

Though a founding member of NATO, Portugal was for many years an isolated country because of its involvement in "colonial" wars in Angola, Guinea, and Mozambique. But Portuguese helicopter pilots fought with great courage and gained fantastic experience during the fighting in Africa.

Portugal has assigned few operational units to NATO, its main service being the provision of the strategically valuable Lajes Air Base in the Azores, which was used by US forces during the Yom Kippur War and the Gulf War.

Portugal has never established an army aviation element. Alouette IIIs that saw service in Africa are still operational with two squadrons that fly 24 SA 316s: the training squadrons Esquadra 111 and Esquadra 552. Both squadrons are based at Tancos.

Twelve Aérospatiale SA 330C Pumas, which are currently undergoing an upgrade, are assigned to Sea/Air Rescue. Some Alouette IIs perform light transport and army assistance duties with optional armament of 70mm rockets or 7.62mm MGs. While Portugal was to have purchased about 50 surplus UH-1Hs from the US, that deal has fallen through.

Order of Battle

Esquadra 552, flying Alouette IIIs, based at Tancos;

Esquadra 751, flying Pumas, based at Montijo;

Esquadra 752, flying Pumas, based at Lajes;

Esquadra 111 (Training), flying Alouette IIIs, based at Tancos.

An SA 330C Puma (the military transport variant of the Puma) belonging to Esquadra 752 based at Lajes sits with its landing gear wheels chocked at the airfield in the Azores. The camouflage pattern it wears is more dramatic than most of the other helicopter camouflage patterns we have seen so far. (Benoit Colin)

This particular SA 330C is one of the Portuguese Pumas that is equipped to perform Search and Rescue (SAR) operations. One of the SAR features is the floatation gear attached to the fairing at the rear of the fuselage. Here the Puma skirts the scenic shore of the Azore Islands on its way to practice a rescue mission. (Benoit Colin)

The powerful four-blade main rotor of this hovering SA 330C Puma kicks up a ring of whitecaps in the water off the Azore Islands as it performs a Search and Rescue exercise. Note the brightly dressed crew member in the open doorway. This Puma is one of five that were converted to the SAR model during the late 1980s. (Benoit Colin)

One of the Portuguese SA 330Cs that is fitted for SAR duty goes through rescue procedures within view of the inhabitants of the white-faced buildings in the background. The SAR Pumas are fitted with the OMERA ORB-31 Hercules radar in their prominent nose radomes. (Benoit Colin)

SPAIN

It was only in July 1965 that the Chief of Staff of the Spanish Army decided to create an independent Army Light Aviation Group. In 1973 the unit was redesignated FAMET (Fuerzas Aeromoviles del Ejército de Tierra, or Army Airmobile Forces). Before then, a few fixed-wing Cessna -16s, 2 Bell -47s and 2 Sikorsky H-19s were used by the Air Force to support the armored division "Brunete".

The first helicopters used by Spain were six Bell UH-1Bs purchased from the USA and eight OH-13s furnished by the Italian firm Agusta-Bell. They entered service in 1966 and were used in a unit called Unidad de Helicopteros XI para Cuerpo de Ejército. But the real beginning of the FAMET was in 1970 with the organization of a staff in charge of studying the aeromobility in the Spanish Army.

At that time 13 OH-58 reconnaissance helicopters, 60 UH-1H assault helicopters and 6 CH-47 transport helicopters were ordered from the USA. Because of mounting tension in the Spanish Sahara near the Moroccan border, Spain quickly acquired three French Alouette IIIs fitted with first generation SS-11 wire-guided missiles. They were delivered in 1974 and were the first armed helicopters used by FAMET. However, it was the German BO 105, manufactured under license in Spain, that was selected by the FAMET for the newly created Batallón de Helicópteros de Ataque (BHELA), the Attack Helicopter Battalion.

In the 1980s the FAMET was restructured. More Chinooks arrived, and Spain ordered 18 Cougars to form a field helicopter regiment.

The FAMET Today

Currently the FAMET is organized into six battalions, a transport unit and a flight school called CEFAMET. The units are as follows:

Batallón de Helicópteros de Ataque I (BHELA I), based at Almagro, which flies BO 105s;

Batallón de Helicópteros de Maniobra II (BHELMA II), based at Bétera, which flies BO 105s and UH-1Hs (which are being exchanged for AS 532UL Cougars);

BHELMA III, based at Agoncillo, which flies BO 105s and UH-1Hs;

BHELMA IV, based at El Copero, which flies Cougars;

BHELMA VI, based at Los Rodeos, which flies UH-1Hs and AB 212s;

Batallón de Helicópteros de Transporte V (BHELTRA V), the heavy transport unit, based at Colmenar, which flies CH-47D Chinooks;

Batallón de Transporte, based at Colmenar, which flies UH-1Hs;

SHEL, in Colmenar, which uses BO 105s, Cougars and UH-1Hs.

CEFAMET, in Colmenar, which uses OH-58s, BO 105s and UH-1Hs.

The total number of helicopters used are:

12 OH-58s, called HR.12B in Spain;

70 BO 105s, called HR/A.15 in Spain (31 Loach, 10 trainers, 29 armed with HOT missiles);

56 UH-1Hs, called HU.10B in Spain;

6 AB 212s, called HU.18 in Spain, used in African territory of Ceuta or Melilla;

18 CH-47Ds, called HT.17 in Spain;

18 AS 532UC Cougars, called HT.21 in Spain, designated to operate in southern Spain.

It was planned that the Chinooks, the UH-1Hs and the BO 105s would be upgraded. Also, Spain will soon select a new combat helicopter (either the Eurocopter Tiger or the AH-64A/D Apache).

A FAMET BO 105GSH, called HR.15, of BHELA I flies near the Sommossierra Pass during the Franco-Spanish exercise "Galia 86". The helicopter has been painted in an overall olive drab paint scheme, which is very effective in the hilly terrain. Note the RH 202 20mm cannon mounted on the centreline of the gun-mounted GSH model.

An unarmed "Loach" BO 105GSH observation helicopter in the service of Spain's FAMET takes off in Andalucia during the exercise "Galia 88". This helicopter performed reconnaissance missions for two HOT-armed BO 105s and provided target designation for them.

Clouds of dust swirl around this BHELA I BO 105GSH as it takes off from a temporary campaign airfield during the "Galia 88" exercise in Andalucia. The gun pod for the Rheinmetall RH 202 20mm cannon is clearly visible here. This gun is used primarily for fire suppression and against soft targets.

The anti-armor version of Spain's BO 105 is designated as HA.15. The HA.15s shown here belong to BHELA I, Spain's attack helicopter battalion. The photo was taken during the exercise "Iberia 93", which was held in the southeast part of France.

Parked on a grassy airfield in France, these HA.15s (BO 105ATHs) from BHELA I take a break in between flights during "Iberia 93". Currently there are 29 of this variant of helicopter in service with the FAMET. The open cabin doors allow the viewer to see the relative size of the interior.

This rear view shows the HOT missile pylons attached to the fuselage of an HA.15 antitank helicopter. When flying an actual mission, the helicopter would carry three missiles on each rack. During such a mission the HA.15 is escorted by an HR.15 fitted with a 20mm cannon.

This photograph, taken from behind an HA.15 (BO 105ATH), shows the helicopter preparing to take on fuel before venturing out on another practice flight during the "Iberia 93" exercise. Apart from its internal fuel tank, this helicopter can carry 705 lb (320 kg) of auxiliary fuel in a cabin tank.

A line-up of Bell UH-1Hs (known as HU.10Bs in Spain) belonging to BHELMA III are positioned on a grassy field near the Sommossierra Pass, a historically significant mountain pass on the way to Madrid. As can be deduced from the surrounding terrain, this unit specializes in mountain flying. Spain intends the Hueys in this battalion to be updated.

The last machines to enter FAMAT service were 18 AS 532UC Cougars. They were attached to BHELMA IV based in southern Spain. Fifteen of them are now shared between a headquarters unit and two transport companies. Here one of the AC 532UCs performs transport duty.

This photo shows a Cougar from BHELMA IV taking part in one of the first of many sorties for exercise "Galia 88". With room for 21 passengers, the Cougar is an ideal transport helicopter. FAMET chose two of this spacious model of helicopter to serve as a type of flying command post.

The Spanish designation for the Cougar is HT.21. Here one of the new HT.21s that was recently received into BHELMA II, seen in its standard FAMET paint scheme of olive drab, prepares to unload a group of soldiers during the exercise “Tramontana 94” that was held in southern Spain.

Descending slowly as it lands, an HT.21 Cougar belonging to BHELMA II participates in field maneuvers during “Tramontana 94”. In an effort to maintain a first-rate light aviation unit, the FAMET decided to replace Hueys in BHELMA II with the more modern Cougar.

Serving its purpose well, an HT.21 Cougar drops off an element of the 5th Spanish Airmobile Brigade in the vicinity of Murcia in the south of Spain during exercise “Tramontana 94”. Spain has found the Cougar to be a versatile and reliable helicopter.

The Spanish Army received its first CH-47C Chinook (HT.17s in Spain) in 1972. This HT.17 is part of BHELTRA V, FAMET's transport unit. Note the sand camouflage scheme it wears. Some Spanish Chinooks take part in exercises held in the Canary Islands or in the desert landscape of Ceuta and Melilla.

BHELTRA V's main role is to provide logistic support for the Spanish Army, especially the Parachute Brigade. HT.17 number "ET 416", which belongs to BHELTRA V, is shown after having just unloaded some French paratroops during the exercise "Galia 88" held in southern Spain.

An HT.17 and a UH-1H (HU.10B), both from the BHELTRA V transport unit, take off from Murcia during the "Tramontana 94" exercise. In the 1980s, Spain announced that some of the CH-47Cs that flew in the FAMET would be upgraded to CH-47Ds. These were delivered to BHELTRA V in the early 1990s.

TURKEY

The country of Turkey is a bridge between Europe and the Middle East . . . and between Islam and Christianity. It is the only Asian member of NATO, and with the problems in the Balkans and the independence of the southern republics of the former Soviet Union, the strategic importance of the country is growing. Turkey has had to build an Army Aviation element to match this growth.

The aviation element of the Turkish Army (Türk Kara Kuvletleri, or TKK) is called the Türk Kara Havaciligi (Turkish Army Aviation). All of Turkey's aircraft are controlled by the Central Army Aviation Establishment at Güvercinlik. The organization of Turkish Army Aviation is as follows:

Hava Taburu (Aviation Battalions), which operate 30 aircraft and helicopters. The battalions are broken down into Hava Boluk (Aviation Companies).

Three battalions form a Hava Alayi (Aviation Regiment), of which three are numbered. A fourth regiment, Aegean Hava Alayi, based at Izmir, supports forces in the Aegean.

The Turkish Army Aviation School (Türk Kara Havacilik Okulu) is also based at Güvercinlik.

Turkey signed a deal in 1993 to purchase 20 AS 532SC Cougar utility helicopters. Thirty more newer generation AS 532US/A2s have been ordered, but will be assembled at Turkish Aerospace at Akinci. The Turkish government has also ordered nineteen Mil Mi-17 utility helicopters.

A Bell AH-1P Cobra attack helicopter of the Türk Kara Kuvvetleri (Turkish Army) prepares to depart on a training mission. The somber paint job on this Cobra is appropriate considering its deadly potential. Note that the armament on the wing stub includes TOW missile launch tubes and a rocket pod that can hold nineteen 2.75-in (70-mm) Folding-Fin Aircraft Rockets (FFARs).

Two AH-1P Cobras belonging to either 2 or 3 Filo (flight) of the Basic Training Battalion at the Turkish Army Aviation School head out on a training flight. The AH-1Ps are recognizable by the early version of the undernose gun turret and the olive drab paint job of the US Army. Although the basic mission of the battalion is the training of pilots, it can be considered an elite reserve unit of the Turkish Army. It saw extensive and successful service against Kurdish rebels during Operation "Iron".

A beautifully camouflaged Bell AH-1W Super Cobra belonging to 1 Filo of the Attack Training Battalion at Güvercinlik. The Super Cobra is a deadly fire support chopper. Mounted in its nose turret is an M179 20mm three-barrel cannon, and on its underwing pylons are two 19-round rocket launchers.

The first AH-1W Super Cobras to be delivered to Turkey arrived in 1990. Located beneath the windscreen of this one is the TOW missile sighting system. Below it is the gun turret that houses the three-barrel cannon that protrudes from under the chin of the Super Cobra. Note the weapons stores on the 10 ft 7-in (3.23-m) stub wing that extends from the fuselage.

As the pilot of this Turkish AH-1W Super Cobra walks passed his machine, he provides a scale to judge the size of this type of helicopter; the Super Cobra has an overall height of 14 ft 2 in (4.32 m). With rotors turning its overall length is 58 ft (17.68 m) even. The diameters of the rotors themselves is 48 ft (14.63 m) for the two-blade main rotor and 9 ft 9 in (2.97 m) for the tail rotor.

This AH-1W Super Cobra is powered by two GE T700-GE-401 turboshafts, each one being rated at 1,625 shp (1,212 kW). They provide the helicopter with a speed at optimum altitude of 173 mph (278 km/h). The three-tone camouflage scheme worn on the Turkish Army Aviation AH-1Ws is well represented here.

This front view illustrates perfectly the narrow profile of the AH-1W Super Cobra, and the vicious appearance of the type when armed with 68mm rocket pods (mounted on the stub wing stores pylons) and the 20mm nose gun. Note the crew's stepped tandem seating configuration.

The crew of a Turkish Army Aviation AH-1W Super Cobra prepares to take off on a training flight. Army pilots are screened during a 12-grading phase before progressing to a two-stage preliminary training course that lasts a total of 30 hours. The crew member here is standing on the open door that provides access to the ammunition bay.

The first rotary-wing aircraft to be used by the Turkish Army was the UH-1H, though they were actually the Italian-built AB 205A model. UH-1Hs were also sent directly from Bell. A total of 156 were eventually dispersed throughout the four Aviation Battalions.

One of the many AB 205As/UH-1Hs that are so popular with the Turkish Army. Despite their popularity, the Hueys are destined to be replaced by the Sikorsky S-70A. This olive drab Huey wears the standard Turkish national marking of a red and white roundel on its tailboom. Note also the Turkish flag on the its tail.

The future of the Turkish Army's transport and assault helicopters is represented by the Sikorsky S-70A Black Hawk (designed as a replacement for the UH-1) and the AS 532 Cougar. The Black Hawk shown here was photographed at Güvercinlik.

The large main rotor of the Black Hawk, which measures 53 ft 8 in (16.36 m) in diameter, is driven by two General Electric T-700-GE-700 turboshafts. They provide the S-70A with a maximum speed of 167 mph (268 km/h). The tail rotor is canted 20 degrees, which offers increased vertical lift capabilities.

This photo illustrates the squat design of the S-70A Black Hawk, which stands 16 ft 10 in (5.13 m) tall. The cockpit seats two pilots, with the captain sitting in the starboard seat. The crew chief is located immediately behind the pilots' position. This Black Hawk is painted much in the same manner as the US Army's UH-60s, bearing no special markings other than the subdued Turkish Army identification on the fuselage.

The S-70A was designed to transport a squad of soldiers into combat as quickly as possible. The Black Hawk's double-windowed cabin doors slide backwards on both sides of the fuselage to allow rapid entry and exit by the troops it is carrying. Note the transparent panel (one of two) at the nose of the helicopter; located on both port and starboard sides of the nose, they provide the pilot with added visibility during landings.

UNITED KINGDOM

Despite its incontestable experience with the utilization of helicopters, Great Britain had until recently never developed a large airmobile unit. The first British heliborne operations were conducted in Malaya during the 1950s with the big two-rotor Bristol Belvedere and the smaller Bristol Sycamore. The Falklands War proved to the world the courage and professionalism of the British helicopter pilots who flew Scouts, Gazelles and Chinooks.

It was only in 1980 that the headquarters staff of the British forces decided to test the concept of airmobile operations with the creation of the 6 Airmobile Brigade, which was the first trial platform of British airmobile operations. This unit was disbanded a few years later, but the British had gained great experience and taken their first step toward airmobility. On the first day of April 1988, a new unit saw the light of day: the 24 Airmobile Brigade, which was an important part of the newly formed Multi-National Division - Central.

At the beginning, the British airmobile unit was comprised exclusively of antitank and defensive units that were used as general reserve for the British Army of the Rhine (BAOR). They were meant to be used in case of war, when they would slow down flank attacks and provide a mobile blockade of the enemy armored columns prior to front stabilization and reinforcement. With the end of the Cold War and the birth of the new NATO Rapid Reaction Corps, the 24 Airmobile Brigade tactics will probably be changed and adapted to a new situation with a more offensive spirit.

The helicopter units that provide machines to the 24 Airmobile Brigade are part of two great commands: the Royal Air Force Support Helicopter Force and the Army Air Corps.

The Army Air Corps

A recent study of the AAC shows it being composed of five separate regiments and a number of independent squadrons. The 1 Regiment is stationed in Germany, while 3 Regiment, 4 Regiment, 7 Regiment and 9 Regiment remained in the UK. Permanent flights are deployed in Belize, Cyprus, Suffield (Canada), Northern Ireland, Falkland Islands, Brunei, and Hong Kong.

The British Army made the decision in 1995 to replace the Lynx helicopters with the AH-64D Longbow Apache antitank helicopters. The Apache was selected in preference to the Eurocopter Tiger, and 67 of them are on order for the AAC. These Apaches will be license-manufactured in the UK by Westland Helicopters.

The Royal Air Force Support Helicopter Force

The principal function of the RAF Support Helicopter Force is, and always has been, the support of the Army requirement for tactical air mobility and logistic support. The machines that serve this force are the Puma, the CH-47 Chinook and the Gazelle. These aircraft are based at RAF Odiham in England and RAF Gütersloh in Germany. This latter base operates to serve the BAOR. Both of these bases have a squadron of Chinooks and one Puma. The No.240 Operational Conversion Unit at Odiham, which was redesignated No. 27 (Reserve) Squadron in 1993, flies both types.

Six Chinooks remain in the Falklands and four Pumas support the Belize garrison in Central America. More recently, the majority of the Pumas and Chinooks were deployed to the Persian Gulf, after which the Chinooks moved to Turkey in support of the relief effort for the Kurdish refugees.

Naturally, the full RAF Support Helicopter Force is involved in the great adventure of the NATO Multi-National Division - Central supporting the 24 Airmobile Brigade and other units. During action it is at the brigade level that the mission of the helicopters is decided, whether it be artillery lift, air assault, antitank duty, recce team drop, or simply a logistics mission.

Normally, an RAF heliborne assault can be divided into five phases:

1) Recce by an AAC Gazelle or an unarmed Lynx to localize the landing zone (LZ);

2) Neutralization and destruction of the enemy's antiaircraft facilities by an armed Lynx or an RAF Harrier or Tornado;

3) First rotation with Puma and

A Gazelle from the Army Air Corps' 9 Regiment is seen during the exercise "Key Flight 89" that was held in northern Germany. "Key Flight 89" was the first exercise for the newly created 24 Airmobile Brigade. Note the smart looking camouflage this Gazelle sports.

British pilots flew Gazelle AH.1s over the snowy countryside of Norway during the exercise "Arrowhead Express 88". This Gazelle was attached to the colonel of the Royal Regiment of Fusiliers during that exercise. It was detached from the Gazelle liaison squadron based in England.

This Gazelle AH.1 had only recently returned from the Persian Gulf when this photo was taken. Note the three white recognition stripes around its tail that were worn by coalition force helicopters. This chopper participated in "Certain Shield 91", the first exercise for the new NATO Multi-National Airmobile Division.

Chinook with combat troops;

4) Second rotation with combat troops and fire support;

5) Third rotation with logistics unit and ammunition.

All of these procedures have been repeated many times, and the British pilots are very professional. An operation of this type can be executed in less than 30 minutes.

The RAF also has a small force of rescue helicopters painted in yellow dispatched to several air bases. These helicopters are mainly the Wessex and Sea King types.

The production of the Westland Lynx came about as a result of an Anglo-French helicopter agreement in 1967. This agreement resulted in the manufacture of a highly maneuverable and versatile machine. The Lynx AH.1 battlefield helicopter, an example of which is seen here, first flew on February 11, 1977.

RAF Support Helicopter Force Order of Battle

No. 7 Squadron, based at Odiham, flying Chinooks;

No. 33 Squadron, based at Odiham, flying Pumas;

NO. 27 (Reserve) Squadron, based at Odiham, using Chinooks and Pumas;

NO. 18 Squadron, based at Laarbruch, flying Chinooks and Pumas;

No. 230 Squadron, based at Aldergrove, flying Pumas.

No. 60 Squadron, based at Benson, flying Wessex.

With the first production EH-101 rolled out in March 1996, delivery of the new 22 EH-101 support helicopters for the RAF will commence in early 1999.

British Helicopters in the Gulf War

On October 24, 1990, it was announced that the RAF and Army Air Corps were preparing to dispatch a support helicopter force to assist the growing army contingent in Saudi Arabia. First in the area near the northern border of Kuwait was the 7th Armoured Brigade, which was backed up by USMC helicopters. When the 4th Mechanized Brigade was also deployed, they needed the RAF to back up the 30,000 troops that just formed the 1st Division.

The first fifteen Puma HC.1s were delivered to Jubail as air freight inside a gigantic USAF C-5 Galaxy from November 2 onward, while three Chinooks were flown out on November 24. More helicopters arrived by sea on January 6.

The RAF helicopters joined with the Royal Navy Sea Kings to form the Support Helicopter Force Middle East, but some of the Chinooks were assigned to Special Forces and repainted with a special camouflage. The Pumas and Chinooks remained close to the troops, leaving Jubail for King Khalid Military City (KKMC) on January 22 and then operating in Iraq with the advancing troops. The Chinooks concentrated on logistic resupply for the division, carrying seven-ton ICO containers, reloads for the Multiple Launch Rocket System and other munitions.

During the Gulf War, RAF pilots used night vision goggles on a regular basis, and the uprated inertial navigation equipment proved to be a godsend in the featureless desert terrain. The RAF helicopters were used to airlift the 3,000 Iraqi prisoners of war and for the reoccupation of the British embassy in Kuwait City.

Army Air Corps Gazelle AH.1s and Lynx AH.7s arrived in Jubail in December, and one month later they were deployed to a forward base north of KKMC. The pilots began immediately to familiarize themselves with the Middle Eastern environment, noticing that it lacked the trees and hills of central Europe that could be used for cover. The issuance of night vision goggles and hastily installed infra-red sighting system for the Lynx allowed the cloak of night to redress the imbalance.

The allied coalition's advance in the land war, which began on February 24, was so swift that ground refueling and rearming parties had difficulty in keeping up with the helicopters as they flew with the leading elements of the UK forces deep into Iraq. TOW-armed Lynx helicopters managed to destroy a large number of tanks and APCs during the brief but intense ground war.

This close-up of a Lynx AH.1 shows the slim design of the helicopter. Spinning above the cabin, attached to a French-made one-piece semi-rigid titanium rotor head, are the Lynx's four rotor blades, which have a diameter of 42 ft (12.8 m). Mounted on the port side of the roof of the cabin can be seen the sight that is used to track a fired TOW missile, which is just one of the type's many weapons options.

These Westland Lynx helicopters are painted in the gray/green camouflage scheme worn by AH.1s of the Army Air Corps. They both belong to the 9 Regiment and are shown here taking part in the "Key Flight 89" exercise held in Germany. The 49 ft 9-in (15.2-m) length and 11 ft 6-in (3.5-m) height of the Lynx is well illustrated in this photo.

An Army Air Corps Lynx AH.1 cruises over the German countryside during the exercise "Key Flight 89". The Lynx derives its power from two Rolls-Royce Gem 2 turboshafts, each of which is rated at 900 shp (671 kW), which are mounted alongside each other atop the upper fuselage.

This interesting overhead view of a Westland Lynx of the 9 Regiment soaring over a body of water in northern Germany demonstrates the versatility of the Army Air Corps' two-tone camouflage scheme. The powerful engines of the AH.1 can take the chopper up to a cruising speed of 161 mph (259 km/h). Note the horizontal stabilizer that extends from the starboard side of the tail.

A close-up of the cabin and nose of a Westland Lynx. The cabin of an AH.1 can seat up to 12 troops in addition to a two-man crew. By removing the seats, room is created to accommodate 2,000 lb (907 kg) of freight. Located on the nose are two blade aerials for the helicopter's UHF/VHF homing radios.

A Lynx AH.1 undergoes some maintenance in the field following its return from the Gulf for the "Certain Shield 91" exercise. This Lynx is fitted with launch tubes for TOW missiles, which is the standard Army armament. The AH.1 can be armed with an amazing variety of weapons including air-to-air missiles, rocket launchers, Minigun pods, and antitank missiles.

This Lynx AH.7, which wears a "UN" recognition insignia, prepares to take off from the air base in Ploce, Croatia during peacekeeping activities in the former Yugoslavia. (Since the adoption of the Dayton Peace Agreement, the AH.7s fly in support of NATO's implementation force, or IFOR). The Lynx belongs to the 3 Regiment of the Army Air Corps and is attached to the Rapid Reaction Corps. Each company in the regiment has 6 Gazelles for liaison and scouting duties and 6 Lynx AH.7s. Note the brilliantly painted tail rotors on the several helicopters.

A 3 Regiment Lynx AH.9, the last version of type to be in the service of the Army Air Corps. The 3 and 4 Regiments attached to the 24 Airmobile Brigade have two antitank companies with Gazelles and Lynx AH.7s and one company with 11 Lynx AH.9s. The Lynx AH.9 is recognizable by its wheeled landing gear.

The Westland Wessex has long served the RAF as a workhorse and search and rescue helicopter. The well known HC.2 first flew in January of 1962. Fifty-four of the veteran helicopters still serve the RAF. One distinguishing feature of the Wessex is the position of the cockpit, which offers the pilot an excellent field of vision.

Two Westland Wessex helicopters wait to load a squad of Belgian paratroops during the exercise "Cold Grouse 95", which may have been the last large exercise in which the Wessex helicopter of No. 60 Squadron will participate. The Wessex's large main rotor, which has a diameter of 56 ft (17 m), is powered by two Roll-Royce Gnome turboshaft engines.

Still performing its duty after all its decades of service, a camouflaged Westland Wessex hangs in the air during "Cold Grouse 95" and shows off its long 65 ft 9-in (20.4-m) fuselage. Operating empty, the Wessex weighs in at 8,304 lb (3,767 kg). The familiar profile of the Wessex is made more memorable by the bulging engine compartment in its nose.

Beginning in 1970, the RAF began receiving forty Puma HC.1 transport helicopters. Here a Puma of No. 230 Squadron of the RAF element based at Gütersloh in Germany waits to go into action during the exercise "Key Flight 89". Note the "tiger head" squadron badge. As a "tiger" unit, No. 230 Squadron is able to participate in the famous Tiger Meet every two years.

A Puma belonging to No. 230 Squadron takes off to conduct a mission during "Key Flight 89" in northern Germany. The Squadron provided helicopter support for the British Army of the Rhine for many years, but it has since returned to the United Kingdom. The Puma bears a certain resemblance to the Sea King HAR.3 flown by No. 78 Squadron, except the Sea King has a much longer fuselage and lacks the Puma's two prominent air intakes.

When based at Gütersloh in Germany, Pumas from No. 230 Squadron provided airlift support for the 1st Battalion of The Royal Green Jackets during the exercise "Key Flight 89". Here a squad of soldiers races to take their place within the spacious cabin of an RAF Puma.

Prior to embarking on a practice exercise as part of "Key Flight 89", pilots and crewmen of RAF Puma HC.1s go over last minute details. The versatility of the Puma makes it a popular support helicopter. Both pilots and aircrews are often called upon to push their skills to the limit while operating these rotary-wing machines.

The picturesque snow-covered landscape of Norway is the setting for support helicopter assistance provided by this Puma HC.1 from No. 33 Squadron. During the exercise "Anchor Express", the Puma prepares to lift a 105mm Light Gun belonging to the Royal Artillery.

Four RAF squadrons are either completely or partially equipped with Puma HC.1 helicopters. One of them is No. 33 Squadron, which is based at Odiham. Here a Puma from No. 33 Squadron, photographed as it just barely begins its lift off, heads out to perform some hauling duties for The Parachute Regiment during the "Blue Lanyard" exercises.

This line-up of RAF Puma HC.1s was photographed at the airport in Jubail in Saudi Arabia just after their arrival. Note the overall desert camouflage paint schemes and the subdued national roundels located on the tailbooms.

Still wearing the remains of desert camouflage that it wore during its service in the Persian Gulf, the three white coalition force recognition stripes being partially obscured due to weathering, a Puma HC.1 from No. 33 Squadron descends to pick up some passengers during the exercise "Ardent Ground 91".

Demonstrating just one of the tasks that makes the HC.1 such a versatile helicopter, this Puma (a weathered veteran of the Gulf War) transports an underslung "Jeep" during air support exercises. The Puma is able to carry 2.5 tons of cargo in this fashion.

Belgian Paracommandos look on in admiration as a Puma HC.1 from the Royal Air Force's No. 33 Squadron makes a delivery of a wheeled vehicle to the "front lines" during "Ardent Ground 91" field exercises. According to reports, the next support helicopter to join the ranks of the RAF will be the EH Industries EH-101.

Gurkhas of the 1/7 Duke of Edinburgh Regiment use a gray and green Royal Air Force Puma HC.1 of No. 240 OCU to hone their skills in battlefield deployment at RAF Odiham. The Puma is able to carry up to 15 fully-equipped soldiers in its cabin.

A Puma flown by No. 240 Operational Conversion Unit (OCU), which was redesignated No. 27 (Reserve) Squadron in 1993, assists in the training of the Gurkhas of 1/7 Duke of Edinburgh Regiment in Odiham. Despite the lack of desert camouflage paint anywhere on this chopper's fuselage, the three white stripes on its tail and underside indicate its status as a veteran of the Gulf War.

This 3/4 view of a Puma HC.1 displays clearly the feature that most easily identifies this type: the two air intakes located above the cabin. In the background is a French Army SA 330 Puma. It is interesting to compare the different paint schemes worn by the two Pumas.

With its cabin packed with what appear to be supplies and material for ground troops, an RAF Puma joins a second Puma in deploying soldiers. Modelers will want to note the almost fluorescent appearance of the squadron insignia against the olive drab paint and the bright red trim painted on the edge of the sliding cabin door.

A Puma from No. 230 (Tiger) Squadron lands to pick up a cabin load of Belgian Paracommandos that are taking part in the exercise "Cold Grouse 95". This joint exercise was held in the vicinity of Nykobing in southern Zealand in Denmark. Note the new two-tone green camouflage paint job that this helicopter recently received.

A Puma HC.1 flies low above wooded countryside on its flight back to its home base following a support operation. The HC.1 has repeatedly demonstrated its worth to the RAF. Not the least of its assets is its troop/cargo carrying ability, being able to haul a maximum payload of 7,055 lb (3,200 kg).

This Puma HC.1 belongs to No. 27 (Reserve) Squadron, the OCU in charge of both the Pumas and Chinooks, is prepared for takeoff from a field near Nykobing during "Cold Grouse 95". Though the Pumas in the RAF are retaining the old designation of "HC.1", those presently in service have been updated with new navigation systems, radar warning receivers, chaff/flare dispensers and infra-red jammers.

This No. 33 Squadron HC.1 is one of the six RAF Pumas that deployed to Croatia with the 24 Airmobile Brigade in August 1995 to support the United Nations forces' attempt to bring some semblance of stability to the former Yugoslavia. Like all the crewmen aboard the Puma, the one seen here in the open doorway is capable of using a 7.62mm machine gun in a hostile environment should the order be given.

An RAF Chinook from No. 7 Squadron - the second Chinook squadron formed - delivers a Light Gun in support of the 5th Brigade during the exercise "Blue Lanyard 88". Along with having an external cargo hook located in the center of its underside, the RAF Chinook HC.1 is equipped with forward and aft hooks that help to steady the underslung load carried by the center hook.

The RAF version of the CH-47 Chinook is the HC.1, and it is truly a support helicopter that can perform. Here a Chinook from No. 7 Squadron lifts a Scimitar tracked vehicle during airborne operations in England.

A Puma HC.1 from No. 33 Squadron, which bears highly visible "UN" identification on its nose and cabin doors, touches down at the airfield at Ploce, Croatia. In the background is parked an RAF CH-47 Chinook, five of which arrived in Croatia along with the other choppers of the 24 Airmobile Brigade.

RAF Chinook are fitted with glassfibre rotor blades, which need an overhaul only every 3,000 miles — half as often as its engines. The Chinook shown here, which belongs to No. 7 Squadron from RAF Odiham, was detached to Denmark to support the UK force participating in the exercise "Bold Grouse 88".

The paint scheme worn by the RAF Chinook HC.1s is well illustrated here. The upper surfaces of the fuselage wear a gray/green camouflage, while the underside is painted black, the edges being feathered in a wave pattern. This Chinook is equipped with a Nitesun searchlight mounted below the nose in between the landing lights.

Ground troops scramble out of the interior of this Chinook HC.1 and down its tail ramp as they go into action. The ramp is operated hydraulically and can be left fully or partially open. This scene was photographed during the exercise "Key Flight 89" that was held in northern Germany.

This RAF Chinook was photographed at RAF Odiham following its return from the Persian Gulf. The crudely applied mottled camouflage scheme of desert pink and black was applied to the Chinook for use in Special Air Service (SAS) night operations. This chopper is fitted with chaff dispensers and an infra-red jammer. Note the hydraulically powered rescue winch mounted above the open main door.

During the exercise "Cold Grouse 95", which was the first operational deployment of NATO Rapid Reaction Corps, eight Chinook HC.2s were sent to Denmark to participate. Here a No. 27(R) Squadron Chinook takes off from Vaerlose, a city near Copenhagen. No. 27(R) Squadron , being the Operational Conversion Unit, uses both Pumas and Chinooks for training.

A Chinook HC.2 belonging to No. 7 Squadron, which is normally based at Odiham, ascends from the airfield at Vaerlose in Denmark during "Cold Grouse 95". When the RAF Chinooks were uprated to HC.2 standards, which roughly correspond to those of the US Army CH-47D, they were repainted in the low IR signature paint scheme of overall dark green seen here. Another sign of the modernization of the RAF helicopters is that their crews are now undergoing intensive training with night vision goggles (NVG).

No. 7 Squadron Chinook HC.2s in UN service at the airfield at Ploce in Croatia. Some of these Chinooks were used on humanitarian missions to deliver food to Serbian refugees in Banja Luka following the Croatian offensive in that region. Currently, Chinooks are being used by IFOR in the former Yugoslavia to deliver troops and equipment, perform reconnaissance, and transport staff.

This Chinook HC.2 of No. 7 Squadron was photographed at Ploce while performing UN peacekeeping duty in Croatia prior to the Dayton Peace Agreement of 1995. Note the drainpipe-like tube that is attached to the helicopter to divert spent shell casings that spew from the machine gun mounted at the main door.

A close-up view of the 7.62mm Minigun (and the crewman that operates it) that is mounted at the main door on the starboard side of this No. 7 Squadron Chinook. Mounted on the port side is an M-60 machine gun. This aggressive HC.2 can be used to deliver or extract Special Forces units or to help rescue downed NATO pilots.

UNITED STATES

Apart from the Germans, the Americans were the first to use vertical-lift rotary-wing aircraft at the end of World War Two. It was in Korea, however, with the use of the Bell 47, the Vertol H-21 and the Sikorsky S-55 that the helicopter established a pedigree. Nevertheless, during and just after the Korean War, the helicopter was considered by the US High Command as merely a utility tool with no real potential as a combat weapon.

Things changed thanks to the initiative of US Army General Carl Hutton who was perceptive enough to conceive and develop the notion of large helicopter formations able to fly and survive in a nuclear battlefield environment. Confirming this view, and despite the opposition of the classical thinking US Army staff, Colonel Jay Vanderpool, who was a disciple of General Hutton, created the Aerial Platoon of Reconnaissance and Experimental Combat in March 1957.

The US Army High Command showed little enthusiasm for the idea of an airmobile unit, considering it as too revolutionary . . . even ridiculous. A unit of this kind was created in Okinawa before in 1962, but when an experimental unit was sent to Viet Nam later that year attitudes changed. In the field the results were impressive. The helicopter showed that is was flexible and well adapted to new as well as classical warfare. The official doctrine toward helicopters finally changed, and under the direction of General Hamilton Howze, the "father of tactical army aviation", the Office of Tactical Mobility began to study the creation of an airmobile division.

A memorable event occurred in the history of helicopter warfare in 1967 when the 1st Cavalry Division, the first airmobile division in the United States Army, successfully engaged in combat during the Battle of Ia Drang in Viet Nam. During the Viet Nam War, the general public became familiar with several helicopters: the Bell UH-1 Huey, the CH-47 Chinook, the OH-6 "Loach", the big S-64 Skycrane, the USMC CH-46, and the CH-53.

Viet Nam also saw the birth of a new kind of helicopter, a real combat helicopter developed specifically for the tasks of war: the AH-1 Cobra. Earlier the French had used an armed S-55 and the 1st Cav modified some Huey into "gunships" armed with M-60 machine guns or an M-134 Minigun, but the Cobra was something unique. It was conceived to escort the assault wave and suppress enemy fire with its 20mm nose gun and its rocket pods. It

The OH-58A/C Kiowa is the US Army's standard observation and light reconnaissance helicopter. It is often used in direct support of battlefield helicopters like the Cobra and the Apache. Here an OH-58A and an AH-1S Cobra from the 2nd Armored Cavalry Regiment are seen on maneuvers during the exercise "Reforger 88".

Developed under the US Army's Armed Helicopter Improvement Program (AHIP), the OH-58D Kiowa Warrior, equipped with a four-blade rotor, is a most effective armed reconnaissance helicopter. Its most distinguishing feature is the mast-mounted observation and target designation system.

This OH-58D Kiowa Warrior and AH-64A Apache are both members of the attack helicopter battalion attached to the 2nd Armored Division ("Hell on Wheels"). The versatile OH-58D works in tandem with the Apache by designating targets for the AH-64A by using the laser spot-tracker on its mast.

is capable of destroying a tank at 3,825.5 yd (3,500 m) with its wire-guided TOW missiles.

A new step in airmobile was taken with the introduction of the AH-64 Apache, which was developed to counter the armored masses of the Warsaw Pact. The Apache is considered to be an airmobile tank killer, able to make long range raids behind enemy lines both day and night. Each division has an Apache battalion, and some independent units can work at the Corps level.

Nowadays the helicopter is omnipresent throughout the US Army, and the Pentagon maintains a fully airmobile division called the 101st Airmobile Assault Division. With thousands in service, the helicopter is one of the best weapons in the United States' arsenal. The USA is doubtless the world's greatest customer for vertical takeoff machines.

Organization

America's largest branch of the armed services is the US Army, which is represented by army HQs. Below the army is the corps. Each corps has a varying number of divisions, and some of the units in a corps include an aviation unit, which includes helicopters. Each division has its own aviation brigade.

During a deployment in Germany during the exercise "Reforger 90", an OH-58D Kiowa Warrior prepares to take off on a reconnaissance mission. The four-blade main rotor, which has a diameter of 35 ft (10. 67 m), has composite blades and a carbon-fiber yoke. Note the blade-like cable cutters that extend from above and below the cockpit.

An OH-58D belonging to the 2nd Armored Division's attack helicopter battalion lands on a German airfield during "Reforger 90". The Kiowa Warrior is powered by a single Allison 250-C30R (military designation T703-AD-700) turboshaft engine, which is rated at 650 shp (485 kW). Note the interesting design of the engine and transmission housing on the top of the fuselage.

The aviation brigade is a mobile force of organic, attached, and supporting army aviation units. They include air assault, attack, reconnaissance, electronic warfare, and support units. Normally an aviation brigade consists of 131 aircraft and has a headquarters and headquarters company (HHC), a cavalry squadron, a command aviation company, an assault helicopter company, and one to two attack helicopter battalion.

The cavalry squadron (Cav) is responsible for reconnaissance and command and control enhancement missions. There is a headquarters and headquarters troop with UH-60, two air troops with AH-1s and OH-58s, and two ground cavalry troops.

The assault helicopter company (AHC) has 18 UH-60s and is the prime air mover of the division. The command aviation company (CAC) provides electronic warfare aircraft and field artillery observers for the division.

The attack helicopter battalion (ATKHB) is employed as a battalion size force and is responsible for antitank and gunship missions. Each battalion has a HHC, three attack companies each with 6 AH-64s, and a maintenance company.

In the Light Divisions, the organization is different and consists of an HHC, a reconnaissance squadron, two assault helicopter companies, and one attack helicopter battalion.

The 101st Air Assault Division has a unique organization structure that is different from the other divisions. The air component is a strong aviation brigade and is composed as follows:

One air cavalry squadron;

Four attack helicopter battalions (AH-64 Apaches);

Two assault helicopter battalions (UH-60s);

One medium helicopter battalion (CH-47 Chinooks);

One command battalion (UH-60s, UH-1s);

One medevac battalion (UH-60s, UH-1s);

One maintenance battalion.

All in all, the 101st Air Assault Division has 317 helicopters in its inventory..

Each armored cavalry regiment has 72 helicopters assigned which includes a regimental aviation squadron with two attack helicopter troops, three air reconnaissance troops, an assault helicopter troop, a headquarters troop and a maintenance company.

This front view of an OH-58D shows the pilot, in the right seat, and his observer, an enlisted man or warrant officer who does not have full flight controls, in the left seat. Their cockpit features two high-resolution multi-function displays, helmet-mounted displays and an up-to-date radar altimeter.

The OH-58D's MMS (mast-mounted sight) can provide targeting and navigation information during both daytime and nighttime flights and in all weather conditions. The McDonnell Douglas creation contains a TV sensor and infra-red system for sighting. The sight is mounted on a special bearing that does not transmit the vibrations from the helicopter.

Lined up against a backdrop of desert that extends to the far distant horizon, several OH-58D Kiowa Warriors wait their turns to conduct reconnaissance missions during the Gulf War. The profile view of the OH-58D at center shows to advantage the design of the 42 ft 2-in (12.85-m) long helicopter.

This OH-58D, which belongs to the 1/82nd "Wolfpack", 82nd Airborne Division, is sitting at an airfield somewhere in the desert of the Persian Gulf. Note how large the main-mounted sight, which is 25.5 in (65 cm) in diameter, appears in comparison to the 12 ft 10-in (3.9-m) tall fuselage of the helicopter.

Displaying a menacing silhouette against the sun-baked sand dunes of the Persian Gulf, an OH-58D from the 82nd Airborne Division speeds through the desert skies on a scouting operation. The Kiowa Warrior has a range of 288 m (463 km) and an endurance time of 2 hours 24 minutes. Its engine can propel the helicopter as fast as 150 mph (241 km/h).

An OH-58D of the 1/1st Cav performs a surveillance mission for the IFOR ground troops somewhere in the skies over Bosnia in 1996. This is one of 16 Kiowa Warriors based in Tuzla that was sent to that region to support the NATO peace implementation force. The rocket pods it carries is one of this type's many weapon options. Other possible armament includes Hellfire antitank and Stinger air-to-air missiles, a 7.62mm Minigun, and a .50 caliber machine gun.

AH-1S Cobras of the 2nd Armored Cavalry Regiment skim the tree tops of the German countryside as they descend into a clearing during the exercise "Reforger 88". The modernized "S" version of the Cobra is the updated variant of the attack helicopter. This view offers a good look at the cockpit's hardened glass windscreen panels, which were designed to reduce attention-getting sun glare.

A well armed AH-1S of the 2nd Armored Cavalry Regiment sets down at the conclusion of a maneuver during "Reforger 88". Fitted to its stub wing store pylons are TOW missile launch tubes, which are supplied in modules of two and are sealed until used. The sight for the TOW missiles is located in the Cobra's nose. Also visible is an M159 19-shot FFAR rocket pod.

One of the tasks assigned to the AH-1S Cobras of the 2nd Armored Cavalry Regiment was to work in cooperation with an advanced team of M-1 Abrams tanks, an element of which is seen here. Note the difference in shade between the dark green paints on the tanks and on the Cobra. The AH-1S is painted with a dark green heat-absorbing paint that reduces the effectiveness of enemy missiles by minimizing infra-red emissions.

This pair of AH-1S Cobra attack helicopters, which belong to the famous 1st Cav, was photographed during the large scale "Reforger 87 Certain Strike" exercise that saw the full deployment of the US III Corps from Fort Hood in Texas to Germany. Visible aft of their engine housings are the straight, extended style engine exhausts that are fitted with black heat suppressers.

An AH-1S Cobra of the 2nd Armored Division revs its engine prior to taking off on an exercise mission during "Certain Centurion 90", which was held in southern Germany. Note the M197 20mm three-barrel rotary cannon that adds to the lethal array of weapons carried by this Cobra.

Striking an aggressive pose, an AH-1S of the 101st Air Assault Division is seen at the very beginning of Operation "Desert Storm" at a resupply depot where the helicopter can take on additional arms and ammunition. The engine that powers this impressive attack helicopter is an Avco Lycoming T53-L-703, which is more reliable than the engines installed in earlier models of the Cobra.

A member of the 1/1st Cav, this AH-1S Cobra displays the standard armament fitted to the attack choppers in Bosnia. Prior to becoming part of the helicopter group supporting IFOR, the Cobras in the former Yugoslavia, which were delivered in C-5 transport aircraft, were assembled at Tasar AFB in Hungary. This close-up view exaggerates the prominence and narrow width of the Cobra's nose.

This formidable machine is the US Army's current antitank and attack helicopter, the AH-64A Apache. The Apache's first appearance in Europe took place during the exercise "Reforger 87 Certain Strike" when the 3rd Battalion of the 6th Air Cavalry Regiment was sent to northern Germany with units of the US III Corps. During the first deployment, the AH-64A suffered some technical problems due to the newness of its technology.

The US Army has a total of 747 AH-64As in its helicopter arsenal. This Apache belongs to the 5th Battalion of the 6th Air Cavalry Regiment. Note how the main landing gear are dramatically angled back away from the undernose cannon. Glistening atop the upper fuselage is an ALQ-144 pulsed heat radiator IRCM (infra-red countermeasures) IR jammer.

Each aviation brigade of the US Army's divisions has 20 Apaches in their attack helicopter battalions. The Apache shown here belongs to the 5th Battalion of the 6th Air Cavalry Regiment. The normal takeoff weight for this bulky aircraft is 14,445 lb (6,552 kg).

Photographed at its base at Wiesbaden, Germany in March 1989, an AH-64A waits its turn to strike terror in the hearts of enemy troops. At the time this picture was taken, the Apache was a frightening adversary for the armored formations of the Warsaw Pact.

An Apache pilot makes final preparations for takeoff prior to embarking on a mission. The engineers of the AH-64A kept the safety of the crew members in mind when they designed the Apache. The cockpit features crash-resistant Kevlar armored seats and lightweight boron shields on each side of the seats for the protection of both pilot and gunner.

Adding to the rather chunky appearance of the AH-64A Apache are two fairings that bulge out like cheeks from each side of the forward fuselage. These fairings actually serve to protect the extensive avionics bays that flank the helicopter. This frontal view of an Apache shows off well the height of the helicopter, which is 14 ft (4.3 m) to the top of the air data sensor that is mounted on the rotor head.

This close-up of the nose of an AH-64A Apache attack helicopter shows the housing that contains TADS (Target Acquisition and Designation Sight) and a night systems sensor scanner. The TADS has a FLIR (forward-looking infra-red), TV camera, a laser tracker and a laser range finder/designator.

Protruding from the mid-section of each side of the Apache's fuselage is a stub wing measuring 17 ft 2 in (5.23 m). This one on the port side of this AH-64A is armed with a pod that contains nineteen 2.75 in (70 mm) Hydra 70 FFAR rockets.

The starboard stub wing of this AH-64A is configured with the standard payload for the Apache: Rockwell Hellfire AGM-114 long-range, laser-guided antitank missiles, which are carried on underwing pylons. The Apache has also been tested with Sidewinder, Stinger and Mistral air-to-air weapons.

Rising from a secluded hideaway in a clearing among the woods, an AH-64A from the 1/3rd Aviation Battalion heads out to rendezvous with other Apaches. The Apache can strike far behind enemy lines day or night and in all kinds of weather. Thanks to the PNVS (Pilot Night-Vision Sensor), Apache pilots can perform nap-of-the-earth night flying.

AH-64A Apaches from the 1/3rd Aviation Battalion gather in a clearing during the exercise "Reforger 90" to refuel and take on much needed ammunition. The stepped seating design of the pilot and co-pilot/gunner positions is well illustrated here.

Sitting in a tree-framed clearing are an OH-58D and an AH-64A , both of which are participants in the antitank exercises that were part of the "Reforger 90" activities. Visible here are the sweptback tips of the blades on the Apache's main rotor, which has a diameter of 48 ft (14.6m).

The cowlings that protrude from each side of the Apache's upper fuselage house the two General Electric T700-GE-701 turboshaft engines that power the Apache. They provide the AH-64A with a never exceed speed of 227 mph (365 km/h) and a maximum vertical rate of climb at sea level of 2,500 ft (762 m) per minute.

In the opening moments of the Gulf War, AH-64A Apaches used 2.75mm rockets, Hellfire missiles, and their 30mm guns to destroy two Iraqi radar bunkers, thereby opening a corridor for strike aircraft to proceed undetected to Baghdad. These long, sleek Apaches have an impressive length of 58 ft (17.7m).

Armed to the teeth, a deadly Apache from the 2nd Armored "Hell on Wheels" Division lifts off to go tank hunting during "Reforger 90". During this exercise, one of the AH-64As from this unit (perhaps the one seen here) was credited with 59 "vehicle kills" in just one night.

Eight AH-64A Apaches form Task Force Normandy from the 101st Air Assault Division's 1st Battalion, 1st Aviation Regiment fired the first allied shots of the Gulf War on January 17, 1991. The Apache's maneuverability makes it a superb weapon in the heart of the battlefield. The helicopter's tailplane, shown here angled downward, assists in the control of the Apache.

The standard color scheme worn by AH-64As is the ubiquitous olive drab with low-visibility black lettering and number. This shot provides a clear view of the armored glass windscreen panels that enclose the cockpit. The round knob protruding from the air intake is the engine transmission gear box.

This AH-64A, taken during Operation "Desert Storm", wears a popular and somewhat weather-beaten shark mouth design on its avionics fairing. Beneath the toothy grin, ground crew members load ammunition for the Apache's M230A-1 30mm Chain Gun, which is mounted under the forward fuselage. This cannon has a withering rate of fire of 625 rounds per minute.

Two Apaches from Task Force Normandy take on a welcome supplies at a makeshift air base. In order to successfully accomplish the attack on the Iraqi radar sites on the first day of the Gulf War, the AH-64As involved in the raid traveled 950 miles (1,528 km) in four days, taking on fuel when necessary at temporary refueling stations.

In December 1995, following the NATO commitment to deploy its implementation force (IFOR) in Bosnia to enforce the peace accords, ten AH-64As flew to the US staging base in Hungary. This is one of those Apaches, a member of the 2/227th "Headhunters", 4th Aviation Brigade. Plans have been made to produce the latest version of the Apache: the AH-64D, which will be equipped with improved engines and Longbow radar mounted on the rotor mast. Re-manufacturing of the first two AH-64As began in early 1996.

US Army AH-64As transferred from Germany are part of the US helicopter force that supports the soldiers of IFOR. This Apache, which carries IFOR lettering, is part of the 2/227th, 4th Aviation Brigade. Proving that it seeks to continually improve the US Army's helicopter force, the Pentagon demonstrated the new RAH-66 Comanche reconnaissance/attack and air combat helicopter early in 1996. It is the first US Army helicopter to have fly-by-wire flight controls, a low observable design, and a five-bladed fantail rotor. It bears a faint resemblance to this AH-64, but has a fenestron tail rotor like the Gazelle.

The UH-1H Iroquois, more commonly (and affectionately) known as the Huey, has been the US Army's principal utility and transport helicopter since the time of the Viet Nam War. This UH-1H from the aviation brigade of the 8th Infantry was photographed in Germany. Note the wire cutter blades on the top and bottom of the cabin.

Although the UH-1H has largely been replaced by the UH-60, the type is still in widespread use, particularly for medical evacuation and general duties. The UH-1H seen here at an airfield in Wiesbaden, Germany belongs to an unidentified unit. Several Apaches are lined up on the runway in the background.

In a formation that dates back to the 1960s, an impressive collection of Bell UH-1Hs from the aviation brigade of the US Army's III Corps demonstrates classic airmobile procedures during the exercise "Certain Shield 91". These Hueys were used to transport VIPs who were in attendance to observe the exercise.

The pilot of this UH-1H, which belongs to the 101st Air Assault Division, stretches his legs while his Huey, already stacked to the ceiling of its cabin with a wide assortment of supplies, is readied for takeoff. This photo illustrates the typical appearance of the Hueys that were utilized so effectively during Operation "Desert Storm".

With its hydraulically-actuated stabilizer angled perpendicular to its fuselage, a UH-60A from V Corps hovers over the ground prior to touching down. The tailplane was designed to assist the Black Hawk in low-speed maneuvers and in assault operations.

The Sikorsky UH-60 Black Hawk is now the US Army's principal utility and transport helicopter. Those pictured here belong to the US Army's V Corps. They were photographed among the forests of southern Germany during the "Reforger 88" exercises that were held there.

Soldiers of the 1st Armored Division scramble to board a UH-60A transport helicopter during the "Reforger 88" exercises. The spacious Black Hawk is 64 ft 10 in (19.76 m) long, and its cabin has the capacity to carry up to 11 troops, 15 if the seats are removed. UH-60s can be used for general duties, air assaults and medevac flights.

The UH-60 Black Hawk saw a lot of service during the brief but hectic Gulf War. This historic photograph, which was taken without official clearance in February 1991 near Rafah just below the Iraqi border, shows a Black Hawk from the 101st Aviation Brigade being refueled and rearmed. It had just come from a combat engagement with Iraqi troops, 400 of which were taken prisoner and evacuated in CH-47s.

During "Desert Storm" UH-60s carried personnel and equipment throughout the entire theater of operations. UH-60V "Dustoff" helicopters, like this one from the 101st Air Assault Division, assisted in the medical evacuation of many casualties. One UH-60 was lost in combat while attempting to rescue a downed F-16 pilot. Another was lost while carrying Special Forces troops deep into Iraq. With 229 UH-60As and 64 UH-60Vs, the Black Hawk was the most numerous of US Army helicopters during the Gulf War.

This unidentified UH-60 Black Hawk was used as a medevac helicopter during the exercise "Dragon Hammer 92". Though it is unusual to see Army helicopters on a US Navy ship, this one flew from and landed on board the USS Shreveport. The sea going Black Hawks were fitted with external fuel tanks, as is shown in this photo.

Fitted to the fuselage of this UH-60A from the 7/227th "Viper", 4th Aviation Brigade depliyed to Bosnia are the pylons of the External Stores Support System (ESSS). They can be used to carry Hellfire antitank missiles as well as external fuel tanks that hold up to 450 gal (1,703 liters) of fuel. Note the angled support struts that provide additional support to the pylons.

This UH-60A is one of the 7/227th "Viper" air ambulance helicopters that is currently providing medevac assistance in Bosnia as part of the IFOR support helicopter force. The color scheme of the Black Hawk is subdued dark green except for the bright red medevac crosses located on the helicopter's nose and external fuel tanks.

Twenty-four EH-60A "Quick Fix" electronic countermeasure versions of the Black Hawk were assigned to serve in the Persian Gulf during the war in that region. The 7/227th "Viper" of the 4th Aviation Brigade, presently serving in the former Yugoslavia, flies this EH-60C "Quick Fix" helicopter as part of NATO's intervention force.

Since 1963, the CH-47 Chinook has been the US Army's standard medium-lift transportation helicopter. Each US Army corps and the 101st Air Assault Division has one regiment of these well known and efficient helicopters. During the Gulf War they logged considerable flight time carrying troops and hauling equipment.

The first flight of a production CH-47D took place in February 1982. In May of that year, the CH-47D entered service with the 101st Air Assault Division. The Chinook seen here, which belongs to the 101st Air Assault Division, was photographed in Saudi Arabia during Operation "Desert Storm".

A CH-47D of the 101st Air Assault Division (noticeably lacking desert camouflage) is seen here parked near the Iraqi border during the Gulf War. Chinooks were employed during the war to transport Iraqi prisoners of war and spearhead the ground offensive launched by the 101st Air Assault Division. One CH-47D was lost in March 1991 just after the ceasefire.

On December 21, 1995, nine CH-47D Chinooks arrived at Taszar Air Force Base in Kaposujlak, Hungary. They represented the advanced element of the US helicopter force going into Bosnia to support IFOR. As the lettering situated on the rear rotor mast attests, the CH-47D featured in this photograph is one of those Chinooks.